Lecture Notes in Computer Science 13626

Founding Editors

Gerhard Goos
Juris Hartmanis

The series Lecture Notes in Computer Science (LNCS), including its subseries Lecture Notes in Artificial Intelligence (LNAI) and Lecture Notes in Bioinformatics (LNBI), has established itself as a medium for the publication of new developments in computer science and information technology research, teaching, and education.

LNCS enjoys close cooperation with the computer science R & D community, the series counts many renowned academics among its volume editors and paper authors, and collaborates with prestigious societies. Its mission is to serve this international community by providing an invaluable service, mainly focused on the publication of conference and workshop proceedings and postproceedings. LNCS commenced publication in 1973.

Vincent Andrearczyk · Valentin Oreiller ·
Mathieu Hatt · Adrien Depeursinge
Editors

Head and Neck Tumor Segmentation and Outcome Prediction

Third Challenge, HECKTOR 2022
Held in Conjunction with MICCAI 2022
Singapore, September 22, 2022
Proceedings

Springer

Editors
Vincent Andrearczyk ⓘ
HES-SO Valais-Wallis University of Applied
Sciences and Arts Western Switzerland
Sierre, Switzerland

Valentin Oreiller ⓘ
HES-SO Valais-Wallis University of Applied
Sciences and Arts Western Switzerland
Sierre, Switzerland

Mathieu Hatt ⓘ
LaTIM, INSERM, University of Brest
Brest, France

Adrien Depeursinge ⓘ
HES-SO Valais-Wallis University of Applied
Sciences and Arts Western Switzerland
Sierre, Switzerland

ISSN 0302-9743   ISSN 1611-3349 (electronic)
Lecture Notes in Computer Science
ISBN 978-3-031-27419-0   ISBN 978-3-031-27420-6 (eBook)
https://doi.org/10.1007/978-3-031-27420-6

This Springer imprint is published by the registered company Springer Nature Switzerland AG
The registered company address is: Gewerbestrasse 11, 6330 Cham, Switzerland

Preface

Head and Neck cancer (H&N) is one of the deadliest cancers. Personalized medicine and treatment planning would benefit from automatic analysis of FluoroDeoxyGlucose (FDG)-Positron Emission Tomography (PET)/Computed Tomography (CT) images, including automatic segmentation of primary tumor and metastatic lymph nodes, as well as prediction of patient outcome. Building upon the success of the first two editions of HECKTOR (2020 and 2021), we proposed major evolutions to the tasks, including: (i) A large increase of dataset, with cases from three new centers, more than doubling the total number of cases; (ii) the extension of the segmentation task to not only the primary tumors, but also the metastatic lymph nodes, with a distinction between the two types of gross tumor volumes; (iii) more challenging and fully-automatic tasks as we neither provided bounding-boxes locating the oropharynx regions, nor the tumoral volumes for the prediction of Recurrence Free Survival (RFS).

We proposed the third edition of the Head and Neck Tumor Segmentation and Outcome Prediction Challenge (HECKTOR 2022[1]) to evaluate and compare the current state-of-the-art methods in medical image segmentation and radiomics. We provided a dataset containing 883 PET/CT images (524 train and 359 test) from nine centers with manually delineated primary tumors and metastatic lymph nodes, clinical data, and RFS time for training cases. Patient clinical data included (with some missing data) center, age, gender, tobacco and alcohol consumption, performance status, Human PapillomaVirus (HPV) status, and treatment (radiotherapy only or additional surgery and/or chemotherapy). The challenge was hosted by the International Conference on Medical Image Computing and Computer Assisted Intervention (MICCAI 2022)[2]. The results were presented at the half-day event on September 22, 2022 as a designated satellite event of the conference.

Among the 30 teams who submitted predicted contours for the segmentation task (Task 1), 24 teams submitted a paper describing their method and results. 22 papers were accepted after a single-blind review process with a minimum of two reviewers per paper. 16 teams submitted their RFS outcome predictions (Task 2) and a total of 12 teams reported their participation in a paper and all were accepted after review. This volume gathers a total of 23 participants' papers together with our overview paper (which underwent the same reviewing process as the participants' papers).

---

[1] https://hecktor.grand-challenge.org/, November 2022.
[2] https://www.miccai2022.org, November 2022.

We thank the committee members, the participants, MICCAI 2022 organizers, the reviewers, and our sponsors Siemens Healthineers Switzerland, Aquilab, Bioemtech and MIM. The papers included in the volume take the Springer Nature policies into account[3].

December 2022

Vincent Andrearczyk
Valentin Oreiller
Mathieu Hatt
Adrien Depeursinge

---

[3] https://www.springernature.com/gp/authors/book-authors-code-of-conduct, November 2022.

Organization

General Chairs

Vincent Andrearczyk   HESSO Valais-Wallis University of Applied Sciences and Arts Western Switzerland, Switzerland

Valentin Oreiller   HESSO Valais-Wallis University of Applied Sciences and Arts Western Switzerland, Switzerland

Mathieu Hatt   University of Brest, France

Adrien Depeursinge   HESSO Valais-Wallis University of Applied Sciences and Arts Western Switzerland, and Lausanne University Hospital (CHUV), Switzerland

Program Committee

Moamen Abobakr   MD Anderson Cancer Center, USA

Azadeh Akhavanallaf   Geneva University Hospital, Switzerland

Panagiotis Balermpas   University Hospital Zurich, Switzerland

Sarah Boughdad   Lausanne University Hospital, Switzerland

Irène Buvat   Paris-Saclay University, France

Leo Capriotti   University of Rouen Normandy, France

Joel Castelli   University of Rennes, France

Catherine Chez Le Rest   Poitiers University Hospital, France

Pierre Decazes   University of Rouen Normandy, France

Ricardo Correia   Lausanne University Hospital, Switzerland

Dina El-Habashy   MD Anderson Cancer Center, USA

Hesham Elhalawani   Cleveland Clinic Foundation, USA

Clifton D. Fuller   MD Anderson Cancer Center, USA

Mario Jreige   Lausanne University Hospital, Switzerland

Yomna Khamis   MD Anderson Cancer Center, USA

Agustina La Greca   University Hospital Zurich, Switzerland

Abdallah Mohamed   MD Anderson Cancer Center, USA

Mohamed Naser   MD Anderson Cancer Center, USA

John O. Prior   Lausanne University Hospital, Switzerland

Su Ruan   University of Rouen Normandy, France

Stephanie Tanadini-Lang	University Hospital Zurich, Switzerland
Olena Tankyevych	Poitiers University Hospital, France
Yazdan Salimi	Geneva University Hospital, Switzerland
Martin Vallières	University of Sherbrooke, Canada
Pierre Vera	University of Rouen Normandy, France
Dimitris Visvikis	University of Brest, France
Kareem Wahid	MD Anderson Cancer Center, USA
Habib Zaidi	Geneva University Hospital, Switzerland

Additional Reviewers

Daniel Abler
Gustavo X. Andrade Miranda
Irène Buvat
Carlos Cardenas
Weronika Celniak
Pierre-Henri Conze
Mara Graziani
Vincent Jaouen
Artur Jurgas
Niccolò Marini
Nataliia Molchanova
Henning Müller
Sebastian Otálora
Vatsal Raina
Kuangyu Shi
Federico Spagnolo
Marek Wodzinski
Alexandre Zwanenburg

Contents

Overview of the HECKTOR Challenge at MICCAI 2022: Automatic Head and Neck Tumor Segmentation and Outcome Prediction in PET/CT

Vincent Andrearczyk[1], Valentin Oreiller[1,2], Moamen Abobakr[3],
Azadeh Akhavanallaf[4], Panagiotis Balermpas[5], Sarah Boughdad[2],
Leo Capriotti[6], Joel Castelli[7,8,9], Catherine Cheze Le Rest[10,11],
Pierre Decazes[6], Ricardo Correia[2], Dina El-Habashy[3], Hesham Elhalawani[12],
Clifton D. Fuller[3], Mario Jreige[2], Yomna Khamis[3], Agustina La Greca[5],
Abdallah Mohamed[3], Mohamed Naser[3], John O. Prior[2], Su Ruan[6],
Stephanie Tanadini-Lang[5], Olena Tankyevych[10,11], Yazdan Salimi[4],
Martin Vallières[13], Pierre Vera[6], Dimitris Visvikis[11], Kareem Wahid[3],
Habib Zaidi[4], Mathieu Hatt[11], and Adrien Depeursinge[1,2(✉)]

[1] Institute of Informatics, HES-SO Valais-Wallis University of Applied Sciences and
Arts Western Switzerland, Sierre, Switzerland
adrien.depeursinge@hevs.ch
[2] Department of Nuclear Medicine and Molecular Imaging, Lausanne University
Hospital (CHUV), Rue du Bugnon 46, 1011 Lausanne, Switzerland
[3] The University of Texas MD Anderson Cancer Center, Houston, USA
[4] Geneva University Hospital, Geneva, Switzerland
[5] University Hospital Zürich, Zurich, Switzerland
[6] Center Henri Becquerel, LITIS Laboratory, University of Rouen Normandy,
Rouen, France
[7] Radiotherapy Department, Cancer Institute Eugène Marquis, Rennes, France
[8] INSERM, U1099, Rennes, France
[9] University of Rennes 1, LTSI, Rennes, France
[10] Centre Hospitalier Universitaire de Poitiers (CHUP), Poitiers, France
[11] LaTIM, INSERM, UMR 1101, Univ Brest, Brest, France
[12] Cleveland Clinic Foundation, Department of Radiation Oncology,
Cleveland, OH, USA
[13] Department of Computer Science, Université de Sherbrooke,
Sherbrooke, QC, Canada

Abstract. This paper presents an overview of the third edition of the
HEad and neCK TumOR segmentation and outcome prediction (HECK-
TOR) challenge, organized as a satellite event of the 25th International
Conference on Medical Image Computing and Computer Assisted Intervention (MICCAI) 2022. The challenge comprises two tasks related to
the automatic analysis of FDG-PET/CT images for patients with Head
and Neck cancer (H&N), focusing on the oropharynx region. *Task 1* is

---

V. Andrearczyk and V. Oreiller—Equal contribution.
M. Hatt and A. Depeursinge—Equal contribution.

© The Author(s), under exclusive license to Springer Nature Switzerland AG 2023
V. Andrearczyk et al. (Eds.): HECKTOR 2022, LNCS 13626, pp. 1–30, 2023.
https://doi.org/10.1007/978-3-031-27420-6_1

the fully automatic segmentation of H&N primary Gross Tumor Volume (GTVp) and metastatic lymph nodes (GTVn) from FDG-PET/CT images. *Task 2* is the fully automatic prediction of Recurrence-Free Survival (RFS) from the same FDG-PET/CT and clinical data. The data were collected from nine centers for a total of 883 cases consisting of FDG-PET/CT images and clinical information, split into 524 training and 359 test cases. The best methods obtained an aggregated Dice Similarity Coefficient (DSC_{agg}) of 0.788 in Task 1, and a Concordance index (C-index) of 0.682 in Task 2.

Keywords: Challenge · Medical imaging · Head and neck cancer · Segmentation · Radiomics · Deep learning · Machine learning

1   Introduction: Research Context

Automatic analysis of multimodal images using machine/deep learning pipelines is of increasing interest. In particular, in the context of oncology, the automation of tumors and lymph nodes delineation can be used for diagnostic tasks (tumor detection), automated staging and quantitative assessment (e.g. lesion volume and total lesion glycolysis), as well as radiotherapy treatment planning and fully automated outcome prediction. This automation presents multiple advantages over manual contouring (faster, more robust and reproducible). Concerning patient-level outcome prediction, multimodal image analysis with machine learning can be used for predictive/prognostic modeling (e.g. response to therapy, prediction of recurrence and overall survival) where image-derived information can be combined with clinical data. These models can be exploited as decision-support tools to improve and personalize patient management.

In this context, the HEad and neCK TumOR segmentation and outcome prediction (HECKTOR) challenge was created in 2020 as a satellite event of MICCAI, with a focus on Head and Neck (H&N) cancer and the use of Positron Tomography Emission / Computed Tomography (PET/CT) images. The first edition of the challenge included a single task, dedicated to the automatic delineation of the primary tumor in combined PET/CT images [6,33]. The second edition (2021) added a second task dedicated to the prediction of Progression-Free Survival (PFS), as well as additional cases from new clinical centers [3]. For the present 2022 (third) edition, the dataset was further expanded (from 425 cases/6 centers to 883 cases/9 centers), and the tasks were updated. For this 2022 edition, Task 1 included the detection and delineation of both the primary tumors and lymph nodes from entire images (no bounding box of the oropharyngeal region was provided as opposed to previous editions), thus achieving a fully automatic segmentation of all pathological targets in the H&N region. Detecting and segmenting both the primary tumor and lymph node volumes opens the avenue to automated TN staging, as well as H&N prognostic radiomics modeling based not only on primary Gross Tumor Volumes (GTVp), but also metastatic lymph nodes (GTVn). The endpoint for Task 2, previously PFS, was change. In the new edition, Recurrence-Free Survival (RFS) was used, and we focused on automatic prediction (no reference contours were provided for the test cases).

While HECKTOR was one of the first challenges to address tumor segmentation in PET/CT images, other challenges are being organized on this topic. In particular, the AutoPET challenge was organized for the first time in 2022 at MICCAI[1]. The objective of the AutoPET challenge was tumor lesion detection and segmentation in whole-body PET/CT [12]. Based on PET images only, a first challenge on tumor segmentation was previously proposed at MICCAI 2016 [16]. The dataset included both simulated and clinical images. Besides challenges, reviews of general automatic tumor segmentation can be found in [17,37].

Whereas an exponentially increasing number of studies were published on oncological outcome prediction based on PET/CT radiomics [18], challenges addressing this type of task are far less popular than the ones focusing on segmentation tasks. Overall, large-scale validation of both tumor segmentation and radiomics based on PET/CT remains insufficiently addressed, highlighting the importance of this third edition of the HECKTOR challenge.

The paper is organized as follows. Section 2 describes the dataset used for each task. Details concerning evaluation metrics, participation and participants' approaches are detailed in Sects. 3 and 4 for Tasks 1 and 2, respectively. The main findings of the challenges are discussed in Sect. 5 while the conclusions of this 2022 edition are summarized in Sect. 6. Appendix 1 contains additional general information and Appendix 2 details PET/CT image acquisitions.

2   Dataset

2.1   Mission of the Challenge

Biomedical Application
The participating algorithms target the following fields of application: diagnosis, prognosis and research. The participating teams' algorithms were designed for either or both image segmentation (i.e., classifying voxels as either primary tumor, metastatic lymph node or background) and RFS prediction (i.e., ranking patients according to a predicted risk of recurrence). The main clinical motivations for these tasks are introduced in Sect. 1.

Cohorts. As suggested in [28], we refer to the patients from whom the image data were acquired as the challenge cohort. The target cohort[2] comprises patients received for initial staging of H&N cancer.

The clinical goals are two-fold; the automatically segmented regions can be used as a basis for (i) treatment planning in radiotherapy, (ii) further investigations to predict clinical outcomes such as overall survival, disease-free survival,

---

[1] https://autopet.grand-challenge.org/, as of November 2022.
[2] The target cohort refers to the subjects from whom the data would be acquired in the final biomedical application. It is mentioned for additional information as suggested in BIAS [28], although all data provided for the challenge are part of the challenge cohort.

response to therapy or tumor aggressiveness. The RFS outcome prediction task does not necessarily have to rely on the output of the segmentation task. In the former case (i), the regions will need to be further refined or extended for optimal dose delivery and control. The challenge cohort[3] includes patients with histologically proven H&N cancer who underwent radiotherapy treatment planning. The data were acquired from nine centers (seven for the training, three for the test, including one center present in both sets) with variations in the scanner manufacturers and acquisition protocols. The data contain PET and CT imaging modalities as well as clinical information including center, age, gender, weight, tobacco and alcohol consumption, performance status, HPV status, and treatment (radiotherapy only or additional chemotherapy and/or surgery). A detailed description of the annotations is reported in Sect. 2.2.

Target Entity. The region from which the image data were acquired (called data origin), varied from the head region only to the whole body, and may vary across modalities. Unlike in previous editions [3,33], we provided the data as acquired, without providing automatic bounding-boxes locating the oropharynx regions [4]. The predictions were evaluated on the entire domain of the CT images.

2.2   Challenge Dataset

Data Source
The data were acquired from nine centers as detailed in Table 1. It consists of FDG-PET/CT images of patients with H&N cancer located in the oropharynx region. The devices and imaging protocols used to acquire the data are described in Table 2. Additional information about image acquisition can be found in Appendix 2.

Training and Test Case Characteristics
The training data comprise 524 cases from seven centers (CHUM, CHUS, CHUP, CHUV, HGJ, HMR and MDA). Only patients with complete responses (i.e. disappearance of all signs of local, regional and distant lesions) after treatment are used for Task 2, i.e. 488 cases. The test data contain 359 cases from two other centers CHB and USZ, and from MDA also present in the training set. Similarly, only patients with complete responses after treatment are used for Task 2, i.e. 339 cases. Examples of PET/CT images of each center are shown in Fig. 1. Each case includes aligned CT and PET images, a mask with values 1 for GTVp, 2 for GTVn, and 0 for background (for the training cases) in the Neuroimaging Informatics Technology Initiative (NIfTI) format, as well as patient information (e.g. age, gender) and center.

Participants who wanted to use additional external data for training were asked to also report results using only the HECKTOR data and discuss differences in the results, but none used external data in this edition.

---

[3] The challenge cohort refers to the subjects from whom the challenge data were acquired.

Table 1. List of the hospital centers in Canada (CA), United States (US), Switzerland (CH) and France (FR) and number of cases, with a total of 524 training and 359 test cases (not all used for task 2, as specified in the rightmost column).

Center	Split	# cases	
		Task 1	Task 2
CHUM: Centre Hospitalier de l'Université de Montréal, Montréal, CA	Train	56	56
CHUS: Centre Hospitalier Universitaire de Sherbooke, Sherbrooke, CA	Train	72	72
HGJ: Hôpital Général Juif, Montréal, CA	Train	55	55
HMR: Hôpital Maisonneuve-Rosemont, Montréal, CA	Train	18	18
CHUP: Centre Hospitalier Universitaire Poitiers, FR	Train	72	44
CHUV: Centre Hospitalier Universitaire Vaudois, CH	Train	53	46
MDA: MD Anderson Cancer Center, US	Train	198	197
Total	Train	524	488
CHB: Centre Henri Becquerel, FR	Test	58	38
MDA: MD Anderson Cancer Center, US	Test	200	200
USZ: UniversitätsSpital Zürich, SW	Test	101	101
Total	Test	359	339

Table 2. List of scanners used in the nine centers. Discovery scanners are from GE Healthcare, Biograph from Siemens, and Gemini from Phillips.

	HGJ	CHUS	HMR	CHUM	CHUV	CHUP	MDA	USZ	CHB	Total
Discovery STE			18	56			133	52		258
Discovery RX							128	24		152
Discovery ST	55						84			139
Biograph 40						72	2			74
Gemini GXL 16		72					1			73
Discovery 690					53		10	8		71
Discovery 710							11		58	69
Discovery HR							2	12		14
Discovery LS							8	3		11
Biograph 64							7			7
Discovery MI							5			5
Biograph 6							1	1		2
Other							2			2
Discovery IQ							1			1
Discovery 600							1			1
Biograph 128								1		1
Biograph 20							1			1
Biograph 16							1			1

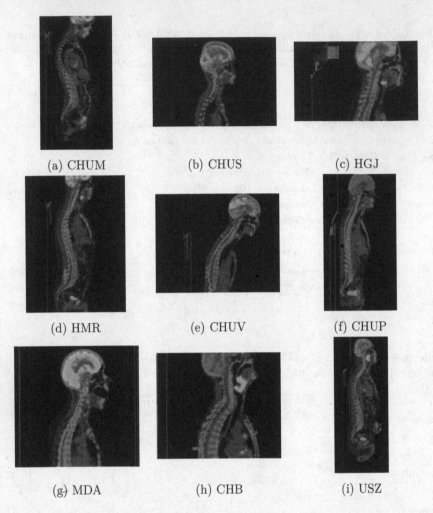

(a) CHUM   (b) CHUS   (c) HGJ

(d) HMR   (e) CHUV   (f) CHUP

(g) MDA   (h) CHB   (i) USZ

Fig. 1. Case examples of 2D sagittal slices of fused PET/CT images from each of the nine centers, showing the variety of fields of view. The CT (grayscale) window in Hounsfield unit is $[-140, 260]$ and the PET window in SUV is $[0, 12]$, represented in a "hot" colormap.

Task 1 - Ground Truth

Original annotations were performed differently depending on the centers.

- Training set CHUV, CHUS, HGJ, HMR: Contours defining the GTVp and GTVn were drawn by an expert radiation oncologist in a radiotherapy treatment planning system. 40% (80 cases) of the training radiotherapy contours were directly drawn on the CT of the PET/CT scan and thereafter used for treatment planning. The remaining 60% (121) of the training radiotherapy contours were drawn on a different CT scan dedicated to treatment planning and were then registered to the FDG-PET/CT scan reference frame using intensity-based free-form deformable registration with the software MIM (MIM software Inc., Cleveland, OH). For the training cases, the original number of annotators is unknown.

- Training set CHUV: The GTVp and GTVn were manually drawn on each FDG-PET/CT by a single expert radiation oncologist.
- Training set CHUP: the metabolic volume of primary tumors was automatically determined with the PET segmentation algorithm Fuzzy Locally Adaptive Bayesian (FLAB) (Hatt et al. 2009) and was then edited and corrected manually by a single expert based on the CT image, for example, to correct cases where the PET-defined delineation included air or non-tumoral tissues in the corresponding CT.
- Training and test sets MDA: Contours available from radiotherapy planning (contoured on the CT image, using a co-registered PET as the secondary image to help physicians visualize the tumor) were refined according to the guidelines mentioned below.
- Test set USZ: The primary tumor was separately segmented in the CT and PET images. The CT segmentation was performed manually. In all cases, two radiation oncologists, both having more than 10 years of experience, were involved in the process. Contours were later post-processed for the presence of metal artifacts to exclude non-tumor-related effects. If a certain tumor slice was affected by any artifacts, the entire tumor contour was erased from that slice. Tumors with more than 50% of volume not suitable for the analysis were not included in the study. Additionally, the voxels outside of soft tissue Hounsfield unit (HU) range (20 HU to 180 HU) were discarded. The tumor in the PET image was auto-segmented using a gradient-based method implemented in MIMVISTA (MIM Software Inc., Cleveland, OH).
- Test set CHB: For each patient, the GTVp and GTVn were manually drawn by using the software PET VCAR (GE Healthcare) on each FDG-PET/CT by senior nuclear medicine physicians using adaptive thresholding with visual control using merged PET and CT information.

Quality controls were performed by experts on all the datasets (training and test) to ensure consistency in ground-truth contours definition. The experts re-annotated them, when necessary, to the real tumoral volume (often smaller than volumes delineated for radiotherapy). A shared cloud environment (MIM Cloud Software Inc.) was used to centralize the contouring task and homogenize annotation software. For cases without original GTVp or GTVn contours for radiotherapy, the experts annotated the cases using PET/CT fusion and N staging information. A guideline was developed by the board of experts for this quality control, reported in the following. Cases with misregistrations between PET and CT were excluded. The annotation guidelines are reported in the following.

Guidelines for primary tumor annotation in PET/CT images. The guidelines were provided to the participants during the challenge.

Oropharyngeal lesions are contoured on PET/CT using information from PET and unenhanced CT acquisitions. The contouring includes the entire edges of the morphologic anomaly as depicted on unenhanced CT (mainly visualized as a mass effect) and the corresponding hypermetabolic volume, using PET acquisition, unenhanced CT and PET/CT fusion visualizations based on automatic co-registration. The contouring excludes the hypermetabolic activity projecting outside the physical limits of the lesion (for example in the lumen of the airway or on the bony structures with no morphologic evidence of local invasion).
Standardized nomenclature per AAPM TG-263: GTVp.
Special situations: Check clinical nodal category to make sure you excluded nearby FDG-avid and/or enlarged lymph nodes (e.g. submandibular, high level II, and retropharyngeal) In case of tonsillar fossa or base of tongue fullness/enlargement without corresponding FDG avidity, please review the clinical datasheet to rule out pre-radiation tonsillectomy or extensive biopsy. If so, this case should be excluded.

Guidelines for nodal metastases tumor annotation in PET/CT images.

Lymph nodes are contoured on PET/CT using information from PET and unenhanced CT acquisitions. The contouring includes the entire edges of the morphologic lymphadenopathy as depicted on unenhanced CT and the corresponding hypermetabolic volume, using PET acquisition, unenhanced CT and PET/CT fusion visualizations based on automatic co-registration for all cervical lymph node levels.
Standardized nomenclature for lymph node ROI: GTVn.
The contouring excludes the hypermetabolic activity projecting outside the physical limits of the lesion (for example on the bordering bony, muscular or vascular structures).

Task 2 - Ground Truth
The patient outcome ground truths for the prediction task were collected in patients' records as registered by clinicians during patient follow-ups. These include locoregional failures and distant metastases. The time t=0 is set to the end date of radiotherapy treatment.

Data Preprocessing Methods
No preprocessing was performed on the images to reflect the diversity of clinical data and to leave full flexibility to the participants. However, we provided various snippets of code to load, crop and resample the data, as well as to evaluate the results on our GitHub repository[4]. This code was provided as a suggestion to help the participants and to maximize transparency (for the evaluation part). The participants were free to use other methods.

Sources of Errors. A source of error originates from the degree of subjectivity in the annotations of the experts [13,33]. Another source of error is the dif-

---

[4] https://github.com/voreille/hecktor, as of November 2022.

ference in the re-annotation between the centers used in HECKTOR 2020 and the one added in HECKTOR 2021/2022. In HECKTOR 2020, the re-annotation was checked by only one expert while for HECKTOR 2021/2022 three experts participated in the re-annotation. Moreover, the softwares used were different.

Finally, another source of error comes from the lack of CT images with a contrast agent for a more accurate delineation of the primary tumor.

Institutional Review Boards
Institutional Review Boards (IRB) of all participating institutions permitted the use of images and clinical data, either fully anonymized or coded, from all cases for research purposes only. More details are provided in Appendix 1.

3   Task 1: Segmentation

3.1   Methods: Reporting of Challenge Design

A summary of the information on the challenge organization is provided in Appendix 1, following the BIAS recommendations.

Assessment Aim. The assessment aim for the segmentation task is to evaluate the feasibility of fully automatic GTVp and GTVn segmentation for H&N cancers via the identification of the most accurate segmentation algorithm.

Assessment Method. The performance is measured by the aggregated Dice Similarity Coefficient (DSC_{agg}) between prediction and manual expert annotations. The DSC_{agg} is computed as followed.

$$\text{DSC}_{\text{agg}} = \frac{2\sum_i |A_i \cap B_i|}{\sum_i |A_i| + |B_i|}, \tag{1}$$

with A_i and B_i respectively the ground truth and predicted segmentation for image i, where i spans the entire test set. This metric was employed in [5].

DSC measures volumetric overlap between segmentation results and annotations. It is a good measure of segmentation for imbalanced segmentation problems, i.e. the region to segment is small as compared to the image size. DSC is commonly used in the evaluation and ranking of segmentation algorithms and particularly tumor segmentation tasks. However, the DSC can be problematic, for instance, for cases without ground truth volume, where a single false negative results in a DSC of 0. GTVn is not present in all images and, if present, there can be more than one volume. The predictions can also include zero, one or more volumes. The proposed DSC_{agg} is well-suited to evaluate this type of task. DSC_{agg} will be computed separately for GTVp and GTVn to account for the smaller number of GTVn. The goal is to identify segmentation methods that perform well on the two types of GTV. A drawback of this metric is that standard deviation (or any statistics) across patients cannot be measured.

3.2   Results: Reporting of Segmentation Task Outcome

Participation. As of September 5, 2022 (submission deadline), the number of registered teams for the challenge (regardless of the tasks) was 121. Each team could submit up to three valid submissions. In order to ensure this limit of submissions, only one participant per team was accepted on grand-challenge and allowed to submit results. By the submission deadline, we had received 67 valid submissions for Task 1, i.e. not accounting for invalid submissions such as format errors. This participation was lower than last year's challenge [3].

In this section, we present the algorithms and results of participants in Task 1 with an accepted paper [1,9,10,21,22,24,25,29,30,32,34–36,38–41,43,45–47, 49]. A full list of results can be seen on the leaderboard[5].

Segmentation: Summary of Participants' Methods. This section summarizes the approaches proposed by all teams for the automatic segmentation of the primary tumor and metastatic lymph nodes (Task 1). The paragraphs are ordered according to the official ranking, starting with the winners of Task 1. Only a brief description of the methods is provided, highlighting the main particularity, without listing the most commonly used training procedures and parameters such as ensembling, losses etc.

In [32], Myronenko et al. used a SegResNet [31] (a 3D U-Net-like architecture with additional auto-encoder and deep supervision) relying on the MONAI[6] platform, adapted to the specificity of the task (e.g. PET/CT, cropping) with the Auto3DSeg[7] system to automate the parameter choice. The main parts of the pipeline involve image normalization, tumor region detection (specific to HECKTOR 2022), isotropic re-sampling, 5-fold cross-validation, and model ensembling. The tumor region detection is based on relative anatomical positions. Random 3D crops are used for training, centered on the foreground classes with probabilities of 0.45 for tumor, 0.45 for lymph nodes and 0.1 for background.

In [41], Sun et al. employed a coarse-to-fine approach with a cascade of multiple networks. 1) The head is first located in CT with a 3D U-Net. 2) A coarse segmentation of GTVp and GTVn regions is performed with a nnU-Net on PET/CT. The ground truth for this step is taken as the center of the GTVp and GTVn. The output is a smaller bounding box centered on the region of interest. 3) Fine segmentation on PET-CT in the finer bounding box is carried out by an ensemble of five nnU-Nets and five nnFormers (trained with cross-validation) using a 3D SE-norm U-Net to generate the final segmentations of GTVp and GTVn.

In [22], Jiang et al. employed an off-the-shelf nnU-NET with simple pre- and post-processing rules. For training, images are cropped around the GTVp. A post-processing outlier removal is based on minimum volume requirements and distance between predicted GTVp and GTVn volumes. Interestingly, an

---

[5] https://hecktor.grand-challenge.org/evaluation/challenge/leaderboard/.

[6] https://github.com/Project-MONAI/MONAI.

[7] https://monai.io/apps/auto3dseg.

integration into a web based platform is proposed for the visualization of the segmentation results, including the segmentation for Organs At Risk (OAR), outside the scope of this challenge.

In [34], Rebaud et al. used a simple nnU-Net-based approach with minor adaptations to the task. In particular, images are resampled to $2 \times 2 \times 2$ mm3, and the training is performed on the entire training set after 5-fold and bagging. Median filtering is used to smooth the resampled masks in the CT resolution.

In [35], Salahuddin et al. proposed a 3D U-Net with a channel-wise attention mechanism, grid-attention gates, carefully designed residual connections and dedicated post-processing to remove outlier volumes on the z-axis. The method is trained using 5-fold cross-validation with extensive data augmentation. Uncommonly, input images are resampled to a non-isotropic $1 \times 1 \times 3$ mm^3 voxel size.

In [45], Wang et al. proposed a base nnU-Net combined with a Transfiner (Vision Transformer, ViT-like model with reduced computation and memory costs) to refine the output, based on the assumption that most segmentation errors occur at the tumor boundaries. The Transfiner treats inputs in a similar manner to a ViT, but uses an octree decomposition of multiple layers of interest to select relevant patches instead of densely patchifying the entire image.

In [46], Wang et al. performed a simple segmentation based on nnU-Net. No region detection is used as pre-processing of the segmentation model. A dense patch-based approach ($128 \times 128 \times 128$) is used with a post-processing based on the distance between GTVp and GTVn to eliminate GTVn volumes that are too far from the GTVp (>150 mm).

In [21], Jain et al. compared several segmentation models: nnU-Net (2D/3D), MNet and SwinU-Net architectures. Images are first resampled to $1 \times 1 \times 3$ mm spacing and then registered altogether using the case CHUM-021 as a reference. Further cropping based on the location of the center of the skull was used for input of all model families.

In [9], Chen et al. built an ensemble of three 3D nnU-Nets trained with different loss functions (Dice + focal loss, Dice + top K loss and cross entropy). The models only take as input the CT images. The PET images are used in a final post-processing step where the U-Nets predictions are penalized based on SUV in the PET.

In [39], Rezaeijo et al. used the following multi-step pipeline. First, an organ localizer module is combined with a 3D U-Net for refined organ segmentation, then a 3D ResU-Net is used to segment GTVp and GTVn. The input of the pipeline is a weighted combination of registered PET and CT images.

In [29], Meng et al. proposed a segmentation network based on a U-Net architecture and a cascaded survival network based on a DenseNet architecture. The two networks are jointly optimized with a segmentation loss and a survival loss. The pipeline jointly learns to predict the survival risk scores of patients and the segmentation masks of tumor regions. It extends the already proposed deep multi-task survival DeepMTS model to a radiomics-enhanced deep multi-task framework.

In [25], Lyu proposes to use a 3D nnU-Net model optimized with the Dice Top-K loss function. An ensemble of the five models obtained from cross-validation is used to produce the GTVp and GTVn segmentation masks.

In [49], Xu et al. applied the nnU-Net framework to the cropped PET/CT images. The PET/CT images are cropped according to the oropharyngeal region which was found relative to the brain detected on the PET images. Two different types of nnU-Nets were used, a "vanilla" nnU-Net and another version that was fine-tuned on the test (referred to as PLL nnU-Net). A combination of Dice and cross-entropy losses was used to train the networks.

In [47], Wang et al. first employed a 2D Retina U-Net [20] to localize the H&N region, followed by a 3D U-Net for the segmentation of GTVp/GTVn.

In [36], Salmanpour et al. trained a Cascade-Net [42] (a cascade of a detection module followed by a segmentation module) on a weighted fusion of PET and CT images.

In [10], Chu et al. used a Swin U-NETR [15] with the encoder pretrained by self-supervision on a large CT dataset. The model is trained with cropped images using the bounding-box extractor provided by the organizers [4].

In [38], Shi et al. used a 3D U-Net-based architecture with inputs of multiple resolutions inputted at different depths in the model. Four resolutions are obtained by randomly cropping the images to a fixed size and resampling them to four dimensions (144^3 and repeatedly halved). The model is trained without a validation set.

In [24], La Greca et al. finetuned two pretrained 3D U-Nets on fused PET-CT images. Both models are pretrained with chest CT images and finetuned for GTVp and GTVn segmentation, respectively. The H&N region is detected semi-automatically (i.e. corrected if necessary) based on the head geometry on the CT image.

In [1], Ahamed et al. proposed to use a 2D ResNet50 pretrained on ImageNet as an encoder in a U-Net-like architecture for slice-wise segmentation, trained without data augmentation. The 3D predictions are obtained on the test set by stacking the 2D predictions.

In [30], Müller et al. performed the localization of the H&N region using an analysis of the PET and CT signals on the z-axis, followed by a simple 3D U-Net for precisely locating the region. Patches were then used in a standard 3D U-Net approach based on the winners of previous editions [19,48], followed by classification for differentiating GTVt and GTVn using Support Vector Machines (SVM).

In [43], Thamawita et al. used a cascade of 2D U-Nets (named TriUnet) in order to merge CT-based predictions and PET-based predictions into a single output prediction.

In [40], Srivastava et al. compared three approaches, two based on explicit multi-scale, previously published by the authors, and one based on Swin UNETR [15], a Swin ViT originally designed for brain tumor segmentation. Despite relatively high performance on validation, the generalization to the test data is not optimal with DSC_{agg} values around 0.5.

Results. The results are reported in Table 3. The results from the participants range from an average DSC_{agg} of 0.48949 to 0.78802. Myronenko et al. [32] obtained the best overall results with an average DSC_{agg} of 0.78802, respectively 0.80066 on the GTVp and 0.77539 on the GTVn. The best GTVn segmentation was obtained by Sun et al. [41] with a DSC_{agg} of 0.77604. Examples of segmentation results are shown in Fig. 2.

Table 3. Results of Task 1. The best out of three possible submissions is reported for each eligible team. Full list of results available at https://hecktor.grand-challenge.org/evaluation/challenge/leaderboard/.

Team	DSC_{agg} GTVp	DSC_{agg} GTVn	mean DSC_{agg}	rank
NVAUTO [32]	**0.80066**	0.77539	**0.78802**	1
SJTU426 [41]	0.77960	**0.77604**	0.77782	2
NeuralRad [22]	0.77485	0.76938	0.77212	3
LITO [34]	0.77700	0.76269	0.76984	4
TheDLab [35]	0.77447	0.75865	0.76656	5
MAIA [45]	0.75738	0.77114	0.76426	6
AIRT [46]	0.76689	0.73392	0.75040	8
AIMers [21]	0.73738	0.73431	0.73584	9
SMIAL [9]	0.68084	0.75098	0.71591	10
Ttest [39]	0.74499	0.68618	0.71559	11
BDAV_USYD [29]	0.76136	0.65927	0.71031	12
junma [25]	0.70906	0.69948	0.70427	13
RokieLab [49]	0.70131	0.70100	0.70115	14
LMU [47]	0.74460	0.65610	0.70035	15
TECVICO Corp [36]	0.74586	0.65069	0.69827	16
RT_UMCG [10]	0.73741	0.65059	0.69400	17
HPCAS [38]	0.69786	0.66730	0.68258	18
ALaGreca [24]	0.72329	0.61341	0.66835	19
Qurit [1]	0.69553	0.57343	0.63448	20
VokCow [30]	0.59424	0.54988	0.57206	21
MLC [43]	0.46587	0.53574	0.50080	22
M&H_lab_NU [40]	0.51342	0.46557	0.48949	23
Average	0.72351	0.68682	0.70517	

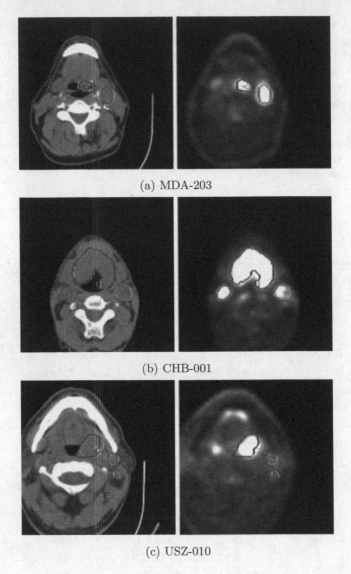

(a) MDA-203

(b) CHB-001

(c) USZ-010

Fig. 2. Examples of results of the winning team (NVAUTO [32]). The automatic segmentation results (light) and ground truth annotations (dark) are displayed on an overlay of 2D slices of CT (left) images and PET (right). GTVn is in red and GTVp in blue. CT are clipped between [−140,260] HU and PET images are between [0,5] SUV.

4   Task 2: Outcome Prediction

The second task of the challenge is the prediction of patient outcome, namely RFS.

4.1   Methods: Reporting of Challenge Design

Due to the connection between the two tasks, this second task was carried out on the same dataset as the first one, exploiting both the available clinical information and the multimodal FDG-PET/CT images. Some patients, however, were not used in the second task (see Table 1) because they did not have a complete response to treatment, which is a pre-requisite for the definition of RFS.

The clinical factors included center, age, gender, weight, tobacco and alcohol consumption, performance status, HPV status, and treatment (radiotherapy only or additional chemotherapy and/or surgery). The information regarding tobacco and alcohol consumption, performance status, HPV status and treatment was available only for some patients. The weight was missing in six training and two test cases, and was estimated to 75 kg to compute the Standard Uptake Values (SUV).

Assessment Aim. The chosen clinical endpoint to predict was RFS, i.e. the time t to reappearance of a lesion or to appearance of new lesions (local, regional or distant), censoring deaths. Only patients with complete responses were considered, and death was censored. In the training set, participants were provided with the survival endpoint to predict, censoring and time-to-event between treatment and event (in days). $t = 0$ was defined as the last day of radiotherapy.

Assessment Method. Challengers were asked to submit a CSV file containing the test patient IDs with the outputs of the model as a predicted risk score anti-concordant with the RFS in days. The performance of the predicted scores was evaluated using the Concordance index (C-index) [14] on the test data. The C-index quantifies the model's ability to provide an accurate ranking of the survival times based on the computed individual risk scores, generalizing the area under the ROC curve. It can account for censored data and represents the global assessment of the model discrimination power. The final ranking was based on the best C-index value obtained on the test set out of the maximum of three submissions per team. The C-index computation is based on the implementation in the Lifelines library [11].

4.2   Results: Reporting of Challenge Outcome

Participation. As mentioned for the first task, the number of registered teams for the challenge (regardless of the tasks) was 121. Each team could submit up to three valid submissions. By the submission deadline, we had received 44 valid submissions for Task 2, i.e. not including invalid submissions e.g. due to format errors. All participants of Task 2 also participated in Task 1.

Outcome Prediction: Summary of Participants' Methods. In this section, we describe the algorithms and results of participants in Task 2 [9, 25, 26, 29, 30, 34–36, 43, 46, 47, 49]. A full list of results can be seen on the leaderboard[8].

---

[8] https://hecktor.grand-challenge.org/evaluation/challenge/leaderboard/.

In [34], Rebaud et al. relied on the Pyradiomics [44] software to extract 93 standard (shape, intensity, textures) radiomics features from the merged GTVp and GTVn delineated volumes (i.e., the result from their task 1 participation) on PET and CT images. In addition to this merged mask, they generated a number of additional masks (re-segmentation with various thresholds, dilation, etc.), which resulted in more than 2400 features per patient. Clinical features were also added, as well as three handcrafted features: the number of tumor masses, the number of lymph nodes, and a binary variable indicating whether the scan was a whole-body scan or included only the H&N region. Each feature was evaluated with the C-index and all pairs of features were also evaluated for their correlation. A novel binary-weighted method was used to assign a binary (-1 / +1) value to each feature, depending on its variation with recurrence time. Finally, the risk was calculated as the mean across all selected feature z-scores weighted by their binary weight. In order to produce a more robust estimate, multiple ensembled models were trained on a random sampling of the training data, also with a randomly selected number of features. A higher number of models led to better performance, and 10^5 models were used on the test set. To evaluate a model on the train set, a two-hundred-fold Monte Carlo cross-validation was used. The ensemble model thus contained three hyperparameters: the number of features randomly drawn for building a model, and the minimum value of C-index and Pearson correlation coefficient threshold to select features among the ones that were randomly drawn. To reduce the risk of overfitting, three bagged models were evaluated in the train and test sets, increasing gradually the number of hyperparameter sets tested, with 10, 100 or 1000 hyperparameter sets, resulting in test C-index values of 0.670, 0.673 and 0.682 respectively.

In [29], Meng et al. proposed an approach similar to the multi-task model trained jointly for both segmentation and prediction task, already proposed in the 2021 edition. The segmentation part is described in Section 3.2. Regarding the outcome prediction task, the model contains a deep learning component, trained on the input PET/CT images, that extracts deep features simultaneously as it generates the segmentation mask. It also contains a standard radiomics component where Pyradiomics features are extracted from the aggregated mask containing the primary tumor and the lymph nodes, as determined by the segmentation part of the pipeline. Finally, clinical variables, deep features and standard radiomics features from the segmentation mask are concatenated in the survival model. The author reported an increased performance in the training set with additional information, with C-index of 0.66, 0.68 and 0.69 relying on clinical variables, standard radiomics and automatic radiomics respectively, whereas the performance of deepMTS only was 0.71, which increased to 0.73 and 0.77 when adding clinical factors then standard radiomics. These three last submissions obtained C-index values of 0.64, 0.65 and 0.68 on the test set. The difference in performance compared with the model ranked 1st is negligible, however, the model is more complex.

In [47], Wang et al. implemented a standard radiomics framework exploiting nnU-Net segmentation, followed by extraction of radiomics features in both PET and CT images (a single mask containing both GTVp and GTVn, using Pyradiomics), followed by feature selection based on univariate analysis, redundancy through correlation, and finally Cox Proportional Hazard (PH) models building through 5-fold CV for each input (clinical, PET, CT) and a combination of the corresponding risk scores. Regarding the clinical variables, missing values were not imputed but instead coded as a third value. The final model combining risk scores from all three inputs (clinical, PET, CT) obtained a C-index of 0.67 in the test set.

In [26], a standard radiomics approach and a deep learning approach were implemented and compared. For the radiomics approach, features were extracted from a delineated volume containing both the GTVp and the GTVn in a single mask, obtained (in both training and test sets) using the segmentation model of task 1. The authors chose to extract shape and intensity metrics from both modalities, and textural features from the CT component only. Features were extracted with Pyradiomics, using a fixed bin width discretization. Radiomics features were then selected and evaluated in a univariate analysis, as well as using correlation to remove redundant ones. Regarding clinical variables, they were selected based on empirical experience as well as univariate analysis and, in the end, only weight and HPV status were retained. The missing HPV values were not imputed but assigned a third category. A Cox PH model using the selected clinical and hand-crafted radiomics features was then trained. The deep-learning model based on a ResNet and the loss function of DeepSurv [23], trained with data augmentation and oversampling, was implemented to also include the clinical features and the hand-crafted radiomics features selected in the radiomics pipeline. Feature selection, model training and validation (for radiomics and deep learning) were all carried out through a 5-fold CV (based on centers, MDA and HMR centers always in the training). The models evaluated on the test set were obtained by averaging the models obtained in each fold. In the test set, a C-index of 0.668 was obtained using the radiomics approach, whereas the DL model (ensemble of DL and radiomics) obtained 0.646, lower than in the validation (>0.75).

In [49], Xu et al. proposed a standard machine learning approach extracting conventional (volume, SUV, TLG, number of nodes etc.) and radiomics features (SERA package [7]) from both PET and CT modalities. The clinical variables were not exploited in the prognostic models. Cox models were trained using either the conventional features alone, the radiomics alone (with ComBat harmonization based on centers), or the combination of conventional and radiomics features (without harmonization). Other combinations were not studied due to the limited number of test submissions. The best result in the test set was obtained by the conventional model relying on Total Lesion Glycolysis (TLG) and the number of nodes features with a C-index of 0.658, whereas the two more complex models led to lower C-index of 0.645 and 0.648.

In [43], Thambawita et al. proposed at first two approaches, one relying on clinical data only, the other combining clinical variables with basic features from the segmentations (volume and z-extent). In both cases, they used Random

Forest. In a third approach, they combined clinical variables with image data using XGBoost. In addition, they estimated the kidney function of the patients and included it as an additional feature, achieving a C-index of 0.656.

In [30], Müller et al. built upon the winning solution of the past challenge ('Deep Fusion V2') that combines a CNN for extracting deep PET and CT features with a Multi-Layer Perceptron (MLP) trained using a multi-class logistic regression loss (MTLR) for survival tasks. It extends this approach by combining the features (deep, shape and intensity features) extracted from multiple image patches (rather than a single patch) via graph convolution. The resulting embeddings are concatenated with clinical information before the final MLP for MTLR loss training. This multi-patch approach uses as inputs the PET/CT fused images cropped at the segmented tumor centroids. It was trained without data augmentation and was compared to Cox PH and Weibull accelerated failure time models relying on clinical variables and basic tumor descriptors only, the original Deep Fusion V2 models as well as combinations of these. Although the new proposed model performed the best in the validation set (C-index 0.75) it failed in the test set (<0.4). The best result in the test set was obtained with the Weibull model (0.64).

In [25], Lyu et al. proposed a method relying on the AutoGluon[9] framework, which consists in an ensemble of 12 models whose outputs are stacked in several successive layers (here 2 layers). The inputs of the models were standard radiomics features calculated from both the PET and CT images using Pyradiomics. Only three clinical variables were considered (gender, age and chemotherapy). This approach obtained a C-index of 0.639 on the test set.

In [46], Wang et al. trained a ResNet model to predict RFS using, as separate channels, the images (PET only, CT only, or PET/CT), with or without the segmentation mask (output of task 1 using a Retina U-Net) through a 3-fold cross-validation. All investigated combinations led to C-index of 0.64–0.70, with the best model obtained using the PET only (0.70). Its prediction performance on the test set (using an averaging of the three models obtained with 3-fold CV) was 0.635.

In [35], Salahuddin et al. focused mainly on the segmentation task (see Sect. 3.2). Nonetheless, they evaluated the prognostic value of some features extracted from the segmentation masks, namely tumor and lymph node largest volumes and number of lymph nodes through a 5-fold cross-validation. A combination of these three features obtained a C-index of 0.627 on the test set.

In [9], Chen et al. extracted standard radiomics features with Pyradiomics from all the individual lesions predicted by the method of Task 1 (compared to most other challengers who chose to consider the whole segmentation mask). The position (center of mass) of each connected component was also concatenated in the vector of radiomics features. Only clinical variables without missing information were used. Prediction of RFS was achieved by training a multiple-instance neural network in order to handle multiple lesions per patient. Amongst various

---

[9] https://auto.gluon.ai/stable/index.html.

training strategies (5-fold CV or the entire training set), the best was using the entire training set, reaching a C-index of 0.619 on the test set.

In [36], Salmanpour et al. extracted deep features from the bottleneck of an auto-encoder fed with PET and CT images fused via a weighted technique. These features were selected with mutual information and fed to a random survival forest trained through a 5-fold CV and grid search, obtaining a C-index of 0.59 on the test set.

Results. The results are reported in Table 4.

Table 4. Results of Task 2. The best out of three possible submissions is reported for each eligible team. Full list of results available at https://hecktor.grand-challenge.org/evaluation/challenge/leaderboard/. The predictions of the MLC team were concordant with the time (prediction of days), instead of a risk score. Their C-index results on the leaderboard were, therefore, < 0.5 and they were ranked last on this task. Other teams made this mistake for their first submission, not reported here because we keep only the best results.

Team	C-index	rank
LITO [34]	**0.68152**	1
BDAV_USYD [29]	0.68084	2
AIRT [46]	0.67257	3
RT_UMCG [26]	0.66834	4
RokieLab [49]	0.65817	5
MLC [43]	0.65598	6
VokCow [30]	0.64081	7
junma [25]	0.63896	8
LMU [47]	0.63536	9
TheDLab [35]	0.6305	10
SMIAL [9]	0.61877	11
TECVICO Corp [36]	0.59042	12
Average	0.64769	

The participants' results range from a C-index of 0.59042 to 0.68152, obtained by Rebaud et al. [34].

5 Discussion: Putting the Results into Context

5.1 Outcomes and Findings

Task 1: Automatic Segmentation of GTVp and GTVn
The participation in this task was slightly lower than in the previous edition [3].

This reduction could be partly due to the limit of three submissions instead of five, as well as the increased difficulty of the task arising from (i) not providing bounding-boxes locating the oropharynx region, and (ii) the need to provide segmentation of both GTVp and GTVn. The quality of the methods and their descriptions, however, was improved. Various successful methods were proposed to detect the oropharynx region prior to inputting data into DL models. Without surprise, the best results were obtained with ensembles of 3D U-Nets with careful design choices for pre and post-processing. The use of transformers increased as compared to the previous editions, without clear benefit on the test performance, but achieving very competitive performance. The winner algorithm (NVAUTO [32]) also performed very well on other MICCAI challenges with adaptations to the tasks.

Besides, a surprisingly high performance was obtained in the segmentation of GTVn (DSC_{agg} =0.687 on average and 0.776 for the best), which one may consider more challenging than the primary tumor due to the large variation in location, size and numbers.

Finally, the reported results are not directly comparable with those of 2021 because of the increased complexity (bounding boxes not provided), the different test set (with results highly influenced by tumor sizes), and the different metrics (DSC_{agg} of GTVp and GTVn in 2022 vs average DSC on GTVp in 2021). If we valuate the GTVp DSC in the winner of 2022 to overcome the metric difference, we obtain a DSC of 0.7056 on the 2022 test set, vs 0.7785 in 2021, highlighting the increased complexity of the present edition and the fact that the algorithms are not optimized solely for GTVp segmentation.

Task 2: RFS Prediction
Similarly to Task 1, the participation was lower than last year's challenge which could be due to the increased complexity of the task. In 2021, we observed a majority of deep-learning based pipelines amongst the top results of the prediction task. Four out of the top 5 results relied on deep learning techniques to extract information from PET/CT images and combine it with clinical data to predict PFS, and only one relied on the extraction of engineered radiomics features, combined through ML algorithms. In the current 2022 edition, most of the best results were obtained through the use of standard radiomics features extraction combined with ML modeling, except the second-ranked team (with almost equal performance as the first rank) that relied on a deep learning setting complemented by standard radiomics features. The winner of the 2021 edition ended up ranked 6th in 2022, building on its previously developed purely DL framework. It should also be emphasized that although the number of training and test cases was more than double the number of the previous edition, the overall performance of the predictive model seemed to reach a plateau around C-index 0.7, not better than in 2021. However, a direct comparison between the two editions is challenging since the data is also more heterogeneous (with additional centers being included), the segmentation task was strongly different and more complex, and the clinical endpoint to predict was slightly different (RFS instead of PFS). Of note, most challengers decided to extract features from a

single mask aggregating the primary tumor and the lymph nodes, which may have biased the prognostic value of some of the features, which may have been more relevant when extracted from each lesion type separately.

Finally, no correlation was observed between the results of tasks 1 and 2, i.e. participants who obtained better results in task 2 were not associated with better results in task 1, with a Pearson correlation of 0.065.

5.2   Limitations of the Challenge

Although important efforts were provided by the multidisciplinary consortium to take attention to details in all research aspects that the HECKTOR challenge is addressing, several limitations remain.

The dataset itself presents several limitations. The contours were drawn based on the PET/(unenhanced)CT fusion, although other methods such as MRI with gadolinium or contrast CT are the gold standard to obtain the true contours for radiation oncology. Since the target clinical application is radiomics, however, the precision of the contours is not as important as for radiotherapy planning. Another limitation is due to the variability in the ground truth annotations. Despite the provided guidelines and the quality checks, some heterogeneity in the annotation methods (e.g. in USZ test collection only, removing lesions with metal artifact n and not in other centers, resegmentation in a given HU range) used by the experts and the experts' profiles were observed. Besides our efforts to unify the contours, this led to a remaining significant source of noise in the labelled data used for training.

Concerning Task 1, the segmentation of GTVp and GTVn, one limitation is the segmentation metric which, despite improving the DSC for the task at hand using the aggregated DSC (see Sect. 3.1), is highly biased towards volume sizes. In the future, we could approach the task as a detection problem, using e.g. the refined Dice proposed by Carass et al. [8].

One limitation of Task 2, the prediction of RFS, was the heterogeneity of the patient cohort in terms of HPV status, age groups and other prognostic factors. To mitigate the impact of this limitation, we provided these clinical parameters to the participants, but missing value rates remained high for some variables (e.g. 36% missing values for HPV status). We also worked on the unification of the RFS definition across all centers (see "Assessment aim" in Sect. 4.1), as we realized that even the definition of RFS itself varied across centers or medical specialties. Besides, the treatment information was relatively homogeneous up to the type, but not how exactly the RT was delivered and the combination with chemo (concomitant or subsequent). This is however realistic regarding clinical practice. While the above-mentioned limitations are commonly admitted in the research community on outcome prediction, we think that focusing on clean populations is key to improving the models' performance when one can afford it in terms of sample size.

Finally, this challenge suffers from other known limitations such as the bias of the Dice with respect to tumor size (large tumors obtain higher Dice scores), and limited ranking robustness [27]. The latter two aspects were investigated for the previous editions [2,33] highlighting a high impact of tumor size on the Dice score and a relatively stable ranking for both segmentation and outcome prediction evaluated with bootstrapping. We expect similar findings for the two tasks of the 2022 edition.

6   Conclusions

This paper presented an overview of the HECKTOR 2022 challenge, dedicated to the automatic analysis of PET/CT images and clinical data of patients with H&N cancer. The tasks proposed in this third edition were (1) Segmentation of primary tumors and metastatic lymph nodes, (2) prediction of patient outcome, namely RFS. The dataset was largely increased in comparison to previous editions, with a total of nine centers and 883 cases. The tasks were also more difficult, in particular with the necessary step of detection of the H&N region prior to further analyses, and the addition of GTVn segmentation to Task 1. Good participation was observed in both tasks, from top research teams across the world proposing a wide variety of methods reported in the 23 quality papers in this volume.

In conclusion, the segmentation results are potentially good enough for clinical use. In future work, we plan to rate the automatic segmentations by experts to assess their quality. Regarding the RFS task, while predictions are better than random, the observed performances suggest that they cannot yet be used clinically to base decisions upon in order to orient treatment options.

Acknowledgments. The organizers thank all the teams for their participation and valuable work. This challenge and the winner prizes were sponsored by Aquilab France, Bioemtech Greece and Siemens Healthineers Switzerland (500€ each, for Task 1, Task 2, and Best Paper). The software used to centralise the annotation and quality control of the GTVp and GTVn regions was MIM (MIM software Inc., Cleveland,OH), which kindly supported the challenge via free licences. This work was also partially supported by the Swiss National Science Foundation (SNSF, grant 205320_179069), the Swiss Personalized Health Network (SPHN, via the IMAGINE and QA4IQI projects) and the RCSO IsNET HECKTOR project.

Appendix 1:   Challenge Information

In this appendix, we list additional important information about the challenge as suggested in the BIAS guidelines [28].

Challenge Name
HEad and neCK TumOR segmentation and outcome prediction challenge (HECKTOR) 2022

Organizing Team
The authors of this paper.

Life Cycle Type
A fixed submission deadline was set for the challenge results.

Challenge Venue and Platform
The challenge is associated with MICCAI 2022. Information on the challenge is available on the website, together with the link to download the data, the submission platform and the leaderboard[10].

Participation Policies

(a) Task 1: Algorithms producing fully-automatic segmentation of the test cases were allowed. Task 2: Algorithms producing fully-automatic RFS risk score prediction of the test cases were allowed.
(b) The data used to train algorithms was not restricted. If using external data (private or public), participants were asked to also report results using only the HECKTOR data.
(c) Members of the organizers' institutes could participate in the challenge but were not eligible for awards.
(d) Task 1: The award was 500 euros, sponsored by Aquilab. Task 2: The award was 500 euros, sponsored by Bioemtech. Best paper award: The award was 500 euros, sponsored by Siemens Healthineers Switzerland.
(e) Policy for results announcement: The results were made available on the grand-challenge leaderboard and the best three results of each task were announced publicly. Once participants submitted their results on the test set to the challenge organizers via the challenge website, they were considered fully vested in the challenge, so that their performance results (without identifying the participant unless permission was granted) became part of any presentations, publications, or subsequent analyses derived from the challenge at the discretion of the organizers.
(f) Publication policy: This overview paper was written by the organizing team's members. The participating teams were encouraged to submit a paper describing their method. The participants can publish their results separately elsewhere when citing the overview paper, and (if so) no embargo will be applied.

Submission Method
Submission instructions are available on the website[11] and are reported in the following.

Task 1: Segmentation outputs should be provided as a single label mask per patient (1 for the predicted GTVp, 2 for GTVn, and 0 for the background) in .nii.gz format. The resolution of this mask should be the same as the original CT resolution. The participants should pay attention to saving NIfTI volumes with the correct pixel spacing and origin with respect to the original reference

---

[10] https://hecktor.grand-challenge.org/.
[11] https://hecktor.grand-challenge.org/Submit/.

frame. The NIfTI files should be named [PatientID].nii.gz, matching the patient names, e.g. CHB-001.nii.gz and placed in a folder. This folder should be zipped before submission. A notebook with a dummy submission example can be found on our github repository[12].

Task 2: Results should be submitted as a CSV file containing the patient ID as "PatientID" and the output of the model (continuous) as "Prediction". An individual output should be anti-concordant with the RFS in days (i.e., the model should output a predicted risk score). If you have a concordant output (e.g. predicted RFS days), you can simply submit your estimate times -1. A notebook with a dummy submission example can be found on our github repository[13].

Participants were allowed three valid submissions per task. The best result was reported in this paper for each task/team.

Challenge Schedule
The schedule of the challenge, including modifications, is reported in the following.

- the release date of the training cases: ~~June 01~~ June 07 2022
- the release date of the test cases: Aug. 01 2022
- the submission date(s): opens Aug. 26 closes ~~Sept. 02~~ Sept. 05 2022 (23:59 UTC-10)
- paper abstract submission deadline: ~~Sept. 02~~ Sept. 05 2022 (23:59 UTC-10)
- full paper submission deadline: Sept. 08 2022 (23:59 UTC-10)
- associated satellite event: Sept. 22 2022

Ethics Approval
Montreal: CHUM, CHUS, HGJ, HMR data (training): The ethics approval was granted by the Research Ethics Committee of McGill University Health Center (Protocol Number: MM-JGH-CR15-50).

CHUV data (training): The ethics approval was obtained from the Commission cantonale (VD) d'éthique de la recherche sur l'être humain (CER-VD) with protocol number: 2018-01513.

CHUP data (training): The fully anonymized data originates from patients who consent to the use of their data for research purposes.

MDA data (training and test): The ethics approval was obtained from the University of Texas MD Anderson Cancer Center Institutional Review Board with protocol number: RCR03-0800.

USZ data (test): The ethics approval was related to the clinical trial NCT01435252 entitled "A Phase II Study In Patients With Advanced Head And Neck Cancer Of Standard Chemoradiation And Add-On Cetuximab".

CHB data (test): The fully anonymized data originates from patients who consent to the use of their data for research purposes.

---

[12] https://github.com/voreille/hecktor/blob/master/notebooks/
example_seg_submission2022.ipynb.

[13] https://github.com/voreille/hecktor/blob/master/notebooks/
example_surv_submission2022.ipynb.

Data Usage Agreement

The participants had to fill out and sign an end-user-agreement, available on the grand-challenge platform, in order to be granted access to the data.

Code Availability

The evaluation software was made available on our github page[14]. The participating teams were encouraged to disclose their code.

Conflict of Interest

No conflict of interest applies. Fundings are specified in the acknowledgments. Only the organizers had access to the test cases' ground truth contours.

Appendix 2:   Image Acquisition Details

HGJ: For the PET portion of the FDG-PET/CT scan, a median of 584 MBq (range: 368–715) was injected intravenously. After a 90-min uptake period of rest, patients were imaged with the PET/CT imaging system (Discovery ST, GE Healthcare). Imaging acquisition of the head and neck was performed using multiple bed positions with a median of 300 s (range: 180–420) per bed position. Attenuation corrected images were reconstructed using an ordered subset expectation maximization (OSEM) iterative algorithm and a span (axial mash) of 5. The FDG-PET slice thickness resolution was 3.27 mm for all patients and the median in-plane resolution was 3.52×3.52 mm^2 (range: 3.52–4.69). For the CT portion of the FDG-PET/CT scan, an energy of 140 kVp with an exposure of 12 mAs was used. The CT slice thickness resolution was 3.75 mm and the median in-plane resolution was 0.98×0.98 mm^2 for all patients.

CHUS: For the PET portion of the FDG-PET/CT scan, a median of 325 MBq (range: 165–517) was injected intravenously. After a 90-min uptake period of rest, patients were imaged with the PET/CT imaging system (Gemini GXL 16, Philips). Imaging acquisition of the head and neck was performed using multiple bed positions with a median of 150 s (range: 120–151) per bed position. Attenuation corrected images were reconstructed using a LOR-RAMLA iterative algorithm. The FDG-PET slice thickness resolution was 4 mm and the median in-plane resolution was 4×4 mm^2 for all patients. For the CT portion of the FDG-PET/CT scan, a median energy of 140 kVp (range: 12–140) with a median exposure of 210 mAs (range: 43–250) was used. The median CT slice thickness resolution was 3 mm (range: 2–5) and the median in-plane resolution was 1.17×1.17 mm^2 (range: 0.68–1.17).

HMR: For the PET portion of the FDG-PET/CT scan, a median of 475 MBq (range: 227–859) was injected intravenously. After a 90-min uptake period of rest, patients were imaged with the PET/CT imaging system (Discovery STE, GE Healthcare). Imaging acquisition of the head and neck was performed using multiple bed positions with a median of 360 s (range: 120–360) per bed position. Attenuation corrected images were reconstructed using an ordered subset

---

[14] https://github.com/voreille/hecktor.

expectation maximization (OSEM) iterative algorithm and a median span (axial mash) of 5 (range: 3–5). The FDG-PET slice thickness resolution was 3.27 mm for all patients and the median in-plane resolution was 3.52×3.52 mm^2 (range: 3.52–5.47). For the CT portion of the FDG-PET/CT scan, a median energy of 140 kVp (range: 120–140) with a median exposure of 11 mAs (range: 5–16) was used. The CT slice thickness resolution was 3.75 mm for all patients and the median in-plane resolution was 0.98×0.98 mm^2 (range: 0.98–1.37).

CHUM: For the PET portion of the FDG-PET/CT scan, a median of 315 MBq (range: 199–3182) was injected intravenously. After a 90-min uptake period of rest, patients were imaged with the PET/CT imaging system (Discovery STE, GE Healthcare). Imaging acquisition of the head and neck was performed using multiple bed positions with a median of 300 s (range: 120–420) per bed position. Attenuation corrected images were reconstructed using an ordered subset expectation maximization (OSEM) iterative algorithm and a median span (axial mash) of 3 (range: 3–5). The median FDG-PET slice thickness resolution was 4 mm (range: 3.27–4) and the median in-plane resolution was 4×4 mm^2 (range: 3.52–5.47). For the CT portion of the FDG-PET/CT scan, a median energy of 120 kVp (range: 120–140) with a median exposure of 350 mAs (range: 5–350) was used. The median CT slice thickness resolution was 1.5 mm (range: 1.5–3.75) and the median in-plane resolution was 0.98×0.98 mm^2 (range: 0.98–1.37).

CHUV: The patients fasted at least 4h before the injection of 4 Mbq/kg of(18F)-FDG (Flucis). Blood glucose levels were checked before the injection of (18F)-FDG. If not contra-indicated, intravenous contrast agents were administered before CT scanning. After a 60-min uptake period of rest, patients were imaged with the PET/CT imaging system (Discovery D690 ToF, GE Healthcare). First, a CT (120 kV, 80 mA, 0.8-s rotation time, slice thickness 3.75 mm) was performed from the base of the skull to the mid-thigh. PET scanning was performed immediately after acquisition of the CT. Images were acquired from the base of the skull to the mid-thigh (3 min/bed position). PET images were reconstructed by using an ordered-subset expectation maximization iterative reconstruction (OSEM) (two iterations, 28 subsets) and an iterative fully 3D (DiscoveryST). CT data were used for attenuation calculation.

CHUP: The acquisition began after 6 h of fasting and 60 ± 5 min after injection of 3 MBq/kg of 18F-FDG (421 ± 98 MBq, range 220–695 MBq), imaged with the PET/CT imaging system (Biograph mCT 40 ToF, Siemens). Non-contrast-enhanced CT images were acquired for attenuation correction (120 kVp, Care Dose® current modulation system) with an in-plane resolution of 0.853×0.853 mm^2 and a 5 mm slice thickness. PET data were acquired using 2.5 min per bed position routine protocol and images were reconstructed using a CT-based attenuation correction and the OSEM-TrueX-TOF algorithm (with time-of-flight and spatial resolution modeling, 3 iterations and 21 subsets, 5 mm 3D Gaussian post-filtering, voxel size $4 \times 4 \times 4$ mm^3).

MDA: For the PET portion of the FDG-PET/CT scan, a median of 401 MBq (range: 327–266) was injected intravenously. After a 90-min uptake period of rest, patients were imaged with the PET/CT imaging system (Multiple hybrid

PET/CT scanner devices). Image acquisition of the head and neck was performed using multiple bed positions with a median of 180 s (range: 90–300) per bed position. Attenuation corrected images were reconstructed using an ordered subset expectation maximization (OSEM) iterative algorithm (2 iterations, 18–24 subsets, 5mm Gaussian filter). The median FDG-PET slice thickness was 3.27 mm (range: 2.99–5) and the median in-plane resolution was 5.46×5.46 mm^2 (range: 2.73×2.73–5.46×5.46). For the CT portion of the FDG-PET/CT scan, a median energy of 120 kVp (range: 100–140) with a median exposure of 185 mAs (range: 10–397) was used. The median CT slice thickness resolution was 3.75mm (range: 2.99–5) and the median in-plane resolution was 0.98×0.98 mm^2 (range: 0.48×0.48–2.734×2.734).

USZ: For PET imaging, an activity of 178–513 MBq was administered intravenously 1h prior to the scan and after the measurement of blood sugar level. Images were acquired with the multiple hybrid PET/CT scanner devices. In the retrospective cohort, 2D or 3D iterative image reconstruction was used, whereas the images of the validation cohort were reconstructed with a 3D algorithm.

CHB: Head and neck PET-CT images were acquired on a GE710 PET/CT device 90 min ($\pm$5 min) after the injection of approximately 3 MBq/kg of FDG. PET and CT acquisition parameters were adapted to the patient's habitus with the patient in the radiotherapy treatment position with a contention mask. For the unenhanced CT portion of the FDG-PET/CT scan, an energy of 120 kVp with an exposure of 25 mAs was used. Attenuation corrected images were reconstructed using an ordered subset expectation maximization (OSEM) iterative algorithm (VPFX, 2 iterations and 23 subsets) and a span (axial mash) of 5. The FDG-PET slice thickness resolution was 3.27 mm for all patients and the median in-plane resolution was 2.73×2.73 mm^2. The CT slice thickness resolution was 2.5 mm and the median in-plane resolution was 0.98×0.98 mm^2 for all patients.

References

1. Ahamed, S., Polson, L., Rahmim, A.: A U-Net convolutional neural network with multiclass Dice loss for automated segmentation of tumors and lymph nodes from head and neck cancer PET/CT images. In: Lecture Notes in Computer Science (LNCS) Challenges (2023)
2. Andrearczyk, V., et al.: Automatic head and neck tumor segmentation and outcome prediction relying on FDG-PET/CT images: findings from the second edition of the HECKTOR challenge. Medical Image Analysis (in review)
3. Andrearczyk, V., et al.: Overview of the HECKTOR challenge at MICCAI 2021: automatic head and neck tumor segmentation and outcome prediction in PET/CT images. In: Andrearczyk, V., Oreiller, V., Hatt, M., Depeursinge, A. (eds.) Head and Neck Tumor Segmentation and Outcome Prediction. HECKTOR 2021. LNCS, vol. 13209, pp. 1–37. Springer, Cham (2022). https://doi.org/10.1007/978-3-030-98253-9_1
4. Andrearczyk, V., Oreiller, V., Depeursinge, A.: Oropharynx detection in PET-CT for tumor segmentation. Irish Mach. Vis. Image Process., 109–112 (2020)

5. Andrearczyk, V., Oreiller, V., Jreige, M., Castelli, J., Prior, J.O., Depeursinge, A.: Segmentation and classification of head and neck nodal metastases and primary tumors in PET/CT. In: 2022 44th Annual International Conference of the IEEE Engineering in Medicine & Biology Society (EMBC), pp. 4731–4735. IEEE (2022)

6. Andrearczyk, V., et al.: Overview of the HECKTOR challenge at MICCAI 2020: automatic head and neck tumor segmentation in PET/CT. In: Andrearczyk, V., Oreiller, V., Depeursinge, A. (eds.) HECKTOR 2020. LNCS, vol. 12603, pp. 1–21. Springer, Cham (2021). https://doi.org/10.1007/978-3-030-67194-5_1

7. Ashrafinia, S.: Quantitative nuclear medicine imaging using advanced image reconstruction and radiomics. Ph.D. thesis, The Johns Hopkins University (2019)

8. Carass, A., et al.: Evaluating white matter lesion segmentations with refined Sørensen-Dice analysis. Sci. Rep. **10**(1), 1–19 (2020)

9. Chen, J., Martel, A.: Head and neck tumor segmentation with 3D UNet and survival prediction with multiple instance neural network. In: Lecture Notes in Computer Science (LNCS) Challenges (2023)

10. Chu, H., et al.: Swin UNETR for tumor and lymph node delineation of multicentre oropharyngeal cancer patients with PET/CT imaging. In: Lecture Notes in Computer Science (LNCS) Challenges (2023)

11. Davidson-Pilon, C.: Lifelines: survival analysis in Python. J. Open Source Softw. **4**(40), 1317 (2019)

12. Gatidis, S., et al.: A whole-body FDG-PET/CT Dataset with manually annotated Tumor Lesions. Sci. Data **9**(1), 1–7 (2022). https://www.nature.com/articles/s41597-022-01718-3

13. Gudi, S., et al.: Interobserver variability in the delineation of gross tumour volume and specified organs-at-risk during IMRT for head and neck cancers and the impact of FDG-PET/CT on such variability at the primary site. J. Med. Imaging Radiat. Sci. **48**(2), 184–192 (2017)

14. Harrell, F.E., Califf, R.M., Pryor, D.B., Lee, K.L., Rosati, R.A.: Evaluating the yield of medical tests. JAMA **247**(18), 2543–2546 (1982)

15. Hatamizadeh, A., Nath, V., Tang, Y., Yang, D., Roth, H.R., Xu, D.: Swin UNETR: swin transformers for semantic segmentation of brain tumors in MRI images. In: Crimi, A., Bakas, S. (eds.) Brainlesion: Glioma, Multiple Sclerosis, Stroke and Traumatic Brain Injuries. BrainLes 2021. LNCS, vol. 12962, pp. 272–284. Springer, Cham (2022). https://doi.org/10.1007/978-3-031-08999-2_22

16. Hatt, M., et al.: The first MICCAI challenge on PET tumor segmentation. Med. Image Anal. **44**, 177–195 (2018)

17. Hatt, M., et al.: Classification and evaluation strategies of auto-segmentation approaches for pet: Report of aapm task group no. 211. Med. Phys. **44**, e1–e42 (2017). https://pubmed.ncbi.nlm.nih.gov/28120467/

18. Hatt, M., et al.: Radiomics in PET/CT: current status and future AI-based evolutions. Seminars Nuclear Med. **51**, 126–133 (2021)

19. Iantsen, A., Visvikis, D., Hatt, M.: Squeeze-and-excitation normalization for automated delineation of head and neck primary tumors in combined PET and CT images. In: Andrearczyk, V., Oreiller, V., Depeursinge, A. (eds.) HECKTOR 2020. LNCS, vol. 12603, pp. 37–43. Springer, Cham (2021). https://doi.org/10.1007/978-3-030-67194-5_4

20. Jaeger, P.F., et al.: Retina U-NET: Embarrassingly simple exploitation of segmentation supervision for medical object detection. In: Machine Learning for Health Workshop, pp. 171–183. PMLR (2020)

21. Jain, A., et al.: Head and neck primary tumor and lymph node auto-segmentation for PET/CT scans. In: Lecture Notes in Computer Science (LNCS) Challenges (2023)
22. Jiang, H., Haimerl, J., Gu, X., Lu, W.: A general web-based platform for automatic delineation of head and neck gross tumor volumes in PET/CT images. In: Lecture Notes in Computer Science (LNCS) Challenges (2023)
23. Katzman, J.L., Shaham, U., Cloninger, A., Bates, J., Jiang, T., Kluger, Y.: Deep-surv: personalized treatment recommender system using a cox proportional hazards deep neural network. BMC Med. Res. Methodol. **18**(1), 1–12 (2018)
24. La Greca Saint-Esteven, A., Motisi, L., Balermpas, P., Tanadini-Lang, S.: A fine-tuned 3D U-net for primary tumor and affected lymph nodes segmentation in fused multimodal images of oropharyngeal cancer. In: Lecture Notes in Computer Science (LNCS) Challenges (2023)
25. Lyu, Q.: Combining nnUNet and AutoML for automatic head and neck tumor segmentation and recurrence-free survival prediction in PET/CT images. In: Lecture Notes in Computer Science (LNCS) Challenges (2023)
26. Ma, B., et al.: Deep learning and radiomics based PET/CT image feature extraction from auto segmented tumor volumes for recurrence-free survival prediction in oropharyngeal cancer patients. In: Lecture Notes in Computer Science (LNCS) Challenges (2023)
27. Maier-Hein, L., et al.: Why rankings of biomedical image analysis competitions should be interpreted with care. Nat. Commun. **9**, 5217 (2018). number: 1 Publisher: Nature Publishing Group
28. Maier-Hein, L., et al.: BIAS: transparent reporting of biomedical image analysis challenges. Med. Image Anal. **66**, 101796 (2020)
29. Meng, M., Bi, L., Feng, D., Kim, J.: Radiomics enhanced deep multi-task learning for outcome prediction in head and neck cancer. In: Lecture Notes in Computer Science (LNCS) Challenges (2023)
30. Muller, A.V.J., Mota, J., Goatman, K., Hoogendoorn, C.: Towards tumour graph learning for survival prediction in head & neck cancer patients. In: Lecture Notes in Computer Science (LNCS) Challenges (2023)
31. Myronenko, A.: 3D MRI brain tumor segmentation using autoencoder regularization. In: Crimi, A., Bakas, S., Kuijf, H., Keyvan, F., Reyes, M., van Walsum, T. (eds.) BrainLes 2018. LNCS, vol. 11384, pp. 311–320. Springer, Cham (2019). https://doi.org/10.1007/978-3-030-11726-9_28
32. Myronenko, A., Siddiquee, M.M.R., Yang, D., He, Y., Xu, D.: Automated head and neck tumor segmentation from 3D PET/CT. In: Lecture Notes in Computer Science (LNCS) Challenges (2023)
33. Oreiller, V., et al.: Head and neck tumor segmentation in PET/CT: the HECKTOR challenge. Med. Image Anal. **77**, 102336 (2022)
34. Rebaud, L., Escobar, T., Khalid, F., Girum, K., Buvat, I.: Simplicity is all you need: out-of-the-box nnUNet followed by binary-weighted radiomic model for segmentation and outcome prediction in head and neck PET/CT. In: Lecture Notes in Computer Science (LNCS) Challenges (2023)
35. Salahuddin, Z., Chen, Y., Zhong, X., Rad, N.M., Woodruff, H., Lambin, P.: HNT-AI: an automatic segmentation framework for head and neck primary tumors and lymph nodes in FDG-PET/CT images. In: Lecture Notes in Computer Science (LNCS) Challenges (2023)
36. Salmanpour, M.R., et al.: Deep learning and machine learning techniques for automated PET/CT segmentation and survival prediction in head and neck cancer. In: Lecture Notes in Computer Science (LNCS) Challenges (2023)

37. Savjani, R.R., Lauria, M., Bose, S., Deng, J., Yuan, Y., Andrearczyk, V.: Automated tumor segmentation in radiotherapy. In: Seminars in Radiation Oncology, vol. 32, pp. 319–329. Elsevier (2022)

38. Shi, Y., Zhang, X., Yan, Y.: Stacking feature maps of multi-scaled medical images in U-Net for 3D head and neck tumor segmentation. In: Lecture Notes in Computer Science (LNCS) Challenges (2023)

39. Rezaeijo, S.M., Harimi, A., Salmanpour, M.R.: Fusion-based automated segmentation in head and neck cancer via advance deep learning techniques. In: Lecture Notes in Computer Science (LNCS) Challenges (2023)

40. Srivastava, A., Jha, D., Aydogan, B., Abazeed, M.E., Bagci, U.: Multi-scale fusion methodologies for head and neck tumor segmentation. In: Lecture Notes in Computer Science (LNCS) Challenges (2023)

41. Sun, X., An, C., Wang, L.: A coarse-to-fine ensembling framework for head and neck tumor and lymph segmentation in CT and PET images. In: Lecture Notes in Computer Science (LNCS) Challenges (2023)

42. Tang, M., Zhang, Z., Cobzas, D., Jagersand, M., Jaremko, J.L.: Segmentation-by-detection: a cascade network for volumetric medical image segmentation. In: 2018 IEEE 15th International Symposium on Biomedical Imaging (ISBI 2018), pp. 1356–1359. IEEE (2018)

43. Thambawita, V., Storas, A., Hicks, S., Halvorsen, P., Riegler, M.: LC at HECKTOR 2022: the effect and importance of training data when analyzing cases of head and neck tumors using machine learning. In: Lecture Notes in Computer Science (LNCS) Challenges (2023)

44. Van Griethuysen, J.J., et al.: Computational radiomics system to decode the radiographic phenotype. Can. Res. **77**(21), e104–e107 (2017)

45. Wang, A., Bai, T., Jiang, S.: Octree boundary transfiner: efficient transformers for tumor segmentation refinement. In: Lecture Notes in Computer Science (LNCS) Challenges (2023)

46. Wang, K., et al.: Recurrence-free survival prediction under the guidance of automatic gross tumor volume segmentation for head and neck cancers. In: Lecture Notes in Computer Science (LNCS) Challenges (2023)

47. Wang, Y., et al.: Head and neck cancer localization with Retina Unet for automated segmentation and time-to-event prognosis from PET/CT images. In: Lecture Notes in Computer Science (LNCS) Challenges (2023)

48. Xie, J., Peng, Y.: The head and neck tumor segmentation using nnU-Net with spatial and channel squeeze & excitation blocks. In: Andrearczyk, V., Oreiller, V., Depeursinge, A. (eds.) HECKTOR 2020. LNCS, vol. 12603, pp. 28–36. Springer, Cham (2021). https://doi.org/10.1007/978-3-030-67194-5_3

49. Xu, H., Li, Y., Zhao, W., Quellec, G., Lu, L., Hatt, M.: Joint nnU-Net and radiomics approaches for segmentation and prognosis of head and neck cancers with PET/CT images. In: Lecture Notes in Computer Science (LNCS) Challenges (2023)

Automated Head and Neck Tumor Segmentation from 3D PET/CT HECKTOR 2022 Challenge Report

Andriy Myronenko$^{(\boxtimes)}$, Md Mahfuzur Rahman Siddiquee, Dong Yang, Yufan He, and Daguang Xu

NVIDIA, Santa Clara, CA, USA
{amyronenko,mdmahfuzurr,dongy,yufanh,daguangx}@nvidia.com

Abstract. Head and neck tumor segmentation challenge (HECKTOR) 2022 offers a platform for researchers to compare their solutions to segmentation of tumors and lymph nodes from 3D CT and PET images. In this work, we describe our solution to HECKTOR 2022 segmentation task. We re-sample all images to a common resolution, crop around head and neck region, and train SegResNet semantic segmentation network from MONAI. We use 5-fold cross validation to select best model checkpoint. The final submission is an ensemble of 15 models from 3 runs. Our solution (team name NVAUTO) achieves the 1st place on the HECKTOR22 challenge leaderboard with an aggregated dice score of 0.78802 (https://hecktor.grand-challenge.org/evaluation/segmentation/leaderboard/). It is implemented with Auto3DSeg (https://monai.io/apps/auto3dseg).

Keywords: HECKTOR22 · MICCAI22 · Segmentation challenge · MONAI · Auto3Dseg · SegResNet · 3D CT · 3D PET

1 Introduction

Head and Neck (H&N) cancer is the fifth most prevalent cancer type globally by incidence rate. Chemotherapy combined with radiotherapy or surgery are standard treatment types, but cancer recurrences occur in almost a half of the cases within the first years after treatments [5]. 3D medical imaging, such as Computed Tomography (CT) and Positron Emission Tomography (PET), provides insights into disease prognosis and treatment planning.

Head and neck tumor segmentation challenge (HECKTOR) provides an opportunity for researchers to develop 3D algorithms for the segmentation of H&N primary tumors (GTVp) in 3D PET/CT scans. HECKTOR 2022 [2,5] is a third edition of the challenge which consists of 883 cases (524 labeled cases were provided for training), each with 3D CT, 3D PET rigidly registered to a common frame, but at different resolutions. The ground truth 3D labels provide dense 3D annotations of 2 structures: gross tumor volumes of the primary

V. Andrearczyk et al. (Eds.): HECKTOR 2022, LNCS 13626, pp. 31–37, 2023.
https://doi.org/10.1007/978-3-031-27420-6_2

tumors (GTVp) and lymph nodes (GTVn). Generally PET images highlight
tumor activity at a lower resolution, whereas CT images provide higher resolution anatomical details. In case of the radiotherapy treatment, the tumor
delineation must be done in the CT coordinate system, which will be used to
calculate the radiation dose to the tumor region. The HECKTOR22 challenge
also includes the second task of outcome prediction, but here we focus solely
on the segmentation task. The data used in this challenge comes from multiple institutions (9 centers in total), including 4 centers in Canada, 2 centers in
Switzerland, 2 centers in France, and 1 center in the United States for a total of
883 patients with annotated GTVp and GTVn [2,5].

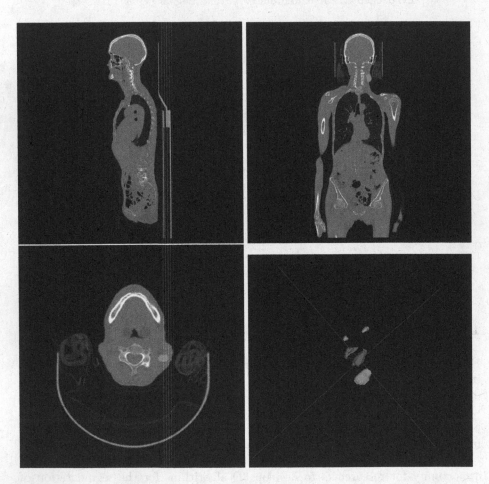

Fig. 1. An example of CT image showing sagital, coronal and axial slices with tumors
(in red) and lymph nodes (in green) mask overlays. A 3D visualization demonstrates
that the tumor consists of 2 components around the neck (in red) and the lymph nodes
region has 3 components annotated (in green). The CT size is $500 \times 500 \times 978$ mm.
(Color figure online)

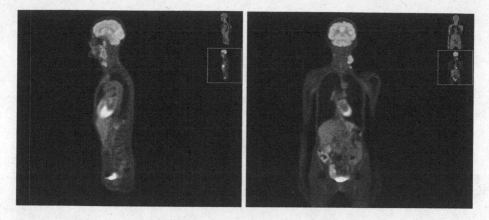

Fig. 2. An example of PET image showing sagital and coronal slices with tumors (in red) and lymph nodes (in green) mask overlays. (Color figure online)

The training dataset with the ground truth labels consists of 524 cases with average 3D CT size of $512 \times 512 \times 200$ voxels at $0.98 \times 0.98 \times 3$ mm average resolution, and with average 3D PET size of $200 \times 200 \times 200$ voxels at $4 \times 4 \times 4$ mm. The CT and PET images where rigidly aligned to a common origin, but remain at different sizes and resolutions. Many cases provided were almost a full body CT/PET pairs. This provides both computational and algorithmic challenge, since the imaging region is as large as $500 \times 500 \times 1000$ mm of the body anatomy, whereas the tumor region covers less then 5% of the input images.

The ground truth labels usually include a single mass of the primary tumor (but in some cases it was absent completely or had two components), and several connected components of the annotated lymph nodes. An example case of CT and the corresponding PET image with ground-truth overlays is shown in Figs. 1 and 2.

2   Materials and Methods

We implemented our approach with MONAI[1] [1], we used Auto3DSeg[2] system to automate most parameter choices. For the main network architecture we used SegResNet[3], which is an encode-decoder based semantic segmentation network based on [4], with deep supervision (see Fig. 3).

Overall, our approach consists of the following steps: data analysis to determine appropriate image normalization parameters and tumor regions, image re-sampling and training of several runs using 5-fold cross-validation, and finally model ensembling.

---

[1] https://github.com/Project-MONAI/MONAI.
[2] https://monai.io/apps/auto3dseg.
[3] https://docs.monai.io/en/stable/networks.html.

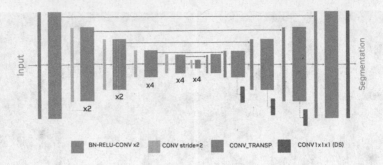

Fig. 3. SegResNet network configuration. The network uses repeated ResNet blocks with batch normalization and deep supervision

Data Preparation. We resample both CT and PET input images to the same size and $1 \times 1 \times 1$ mm isotropic resolution, and crop an approximate region around head and neck. The steps to crop an approximate H&N region are basic and based on a relative anatomy position within the PET/CT images:

- Detect the top of the head (top of the bounding box), based on a simple PET thresholding
- Detect the H&N center-line (xy coordinate) based on the average foreground of top slices
- Crop the bounding box of $200 \times 200 \times 310$ mm centered on the center-line.

This simple approach had 100% success rate on the training set to cover the H&N region fully. We contemplated a more sophisticated approach based on deep-learning, but it was not necessary in this case.

Cropping the approximate region is the first step both during training and during inference. During training it significantly reduces the input image size (e.g. from $500 \times 500 \times 900$ to $200 \times 200 \times 310$ voxels), which speeds up training and avoids unnecessary network strain to differentiate other anatomies (e.g. in abdominal region).

Data Normalization. We re-scale input CT image intensity from a predefined range to 0..1 interval, as determined by data analysis to include the intensity pattern variations within the foreground regions, followed by a sigmoid. For PET image, we normalize it to zero mean and standard deviation one, followed by a sigmoid. Sigmoid function here is used as an alternative to hard intensity clamping. After normalization, input images are concatenated to form a 2 channel input image.

Model. For the model, we used the encoder-decoder semantic segmentation model SegResNet from MONAI based on [4] with deep supervision. The encoder part uses ResNet [3] blocks, and includes 6 stages of 1, 2, 2, 4, 4, 4 blocks respectively. We follow a common CNN approach to downsize image dimensions by 2 progressively and simultaneously increase feature size by 2. All convolutions are $3 \times 3 \times 3$ with an initial number of filters equal to 32. The encoder is trained with $192 \times 192 \times 192$ input region. The decoder structure is similar to the encoder one, but with a single block per each spatial level. Each decoder level begins with upsizing with transposed convolution: reducing the number of features by a factor of 2 and doubling the spatial dimension, followed by the addition of encoder output of the equivalent spatial level. The end of the decoder has the same spatial size as the original image, and the number of features equal to the initial input feature size, followed by a $1 \times 1 \times 1$ convolution into 3 channels and a softmax (a background and two foreground classes).

2.1   Training Method

Dataset. We use the HECKTOR22 dataset [2,5] . We randomly split the entire dataset into 5 folds and trained a model for each fold. We did not use any additional data or pre-trained models, and we did not use any of the meta-data information (such patients gender or age) provided by the organizers.

Cropping. We crop a random patch of $192 \times 192 \times 192$ voxels from the H&N extracted area centered on the foreground classes with probabilities of 0.45 for tumor and 0.45 for lymph nodes (and 0.1 for background).

Augmentations. We use spatial augmentation for both CT and PET images, including random affine and flip in all axes. We also use intensity augmentations only for CT image, including random intensity scale, shift, noise and blurring.

Loss. We use the combined Dice + CrossEntropy loss. The same loss is summed over all deep-supervision sublevels:

$$Loss = \sum_{i=0}^{4} \frac{1}{2^i} Loss(pred, target^{\downarrow}) \tag{1}$$

where the weight $\frac{1}{2^i}$ is smaller for each sublevel (smaller image size) i. The target labels are downsized (if necessary) to match the corresponding output size using nearest neighbor interpolation

2.2   Optimization

We use the AdamW optimizer with an initial learning rate of $2e^{-4}$ and decrease it to zero at the end of the final epoch using the Cosine annealing scheduler. All the models were trained for 300 epochs with deep supervision. We use batch size of 1 per GPU, and train on 8 GPUs 16 Gb NVIDIA V100 DGX machine (which is equivalent to batch size of 8). We use weight decay regularization of $1e^{-5}$.

3 Results

Based on our data splits, a single run 5-folds cross-validation results are shown in Table 1. On average, we achieve 0.7989 cross-validation performance in terms of aggregated Dice metric.

Table 1. Aggregated dice metric using 5-fold cross-validation.

Fold 1	Fold 2	Fold 3	Fold 4	Fold 5	Average
0.7933	0.7862	0.7816	0.8275	0.8059	0.7989

For the final submission we use 15 models total (consisting from best models from 5-fold cross validation, repeating it 3 times). The challenge allowed only 3 submissions total, and required to submit dense prediction masks for 359 cases (saved in CT size/resolution). Our results are in Table 2. All 3 of our submission are the top 3 submissions on HECTOR22 challenge leaderboard[4].

Table 2. Our submission results on HECKTOR22 leaderboard.

Submission	Note	Tumor	Lymph nodes	Total
One	Ensemble mean	0.78797	0.77468	0.78133
Two	Ensemble + TTA	0.80066	0.77539	**0.78802**
Three	+post processing	0.80066	0.77199	0.78632

The three submissions we did are:

- One - a simple mean ensemble of all models.
- Two - we use Test Time Augmentation (TTA) using axis flips (8 flips total) for each model prediction, which resulted in the best performance.
- Three - we attempted to do post-processing on the lymph nodes class based on the submission "Two", by removing small connected components and components with low PET values. Ultimately this heuristic reduced the lymph node accuracy, and was not helpful.

4 Conclusion

In conclusion, in this work, we describe our solution to HECKTOR22 challenge (NVAUTO team). Our automated solution is implemented with MONAI (See footnote 1) and Auto3DSeg (See footnote 2). We achieve the 1st place in the HECKTOR22 challenge segmentation task (See footnote 1).

---

[4] https://hecktor.grand-challenge.org/evaluation/segmentation/leaderboard/.

References

1. Project-monai/monai. https://doi.org/10.5281/zenodo.5083813
2. Andrearczyk, V., et al.: Overview of the HECKTOR challenge at MICCAI 2022: automatic head and neck tumor segmentation and outcome prediction in PET/CT (2023). https://arxiv.org/abs/2201.04138
3. He, Kaiming, Zhang, Xiangyu, Ren, Shaoqing, Sun, Jian: Identity mappings in deep residual networks. In: Leibe, Bastian, Matas, Jiri, Sebe, Nicu, Welling, Max (eds.) ECCV 2016. LNCS, vol. 9908, pp. 630–645. Springer, Cham (2016). https://doi.org/10.1007/978-3-319-46493-0_38
4. Myronenko, Andriy: 3D MRI brain tumor segmentation using autoencoder regularization. In: Crimi, Alessandro, Bakas, Spyridon, Kuijf, Hugo, Keyvan, Farahani, Reyes, Mauricio, van Walsum, Theo (eds.) BrainLes 2018. LNCS, vol. 11384, pp. 311–320. Springer, Cham (2019). https://doi.org/10.1007/978-3-030-11726-9_28
5. Oreiller, V., et al.: Head and neck tumor segmentation in PET/CT: the HECKTOR challenge. Med. Image Anal. 77, 102336 (2022)

A Coarse-to-Fine Ensembling Framework for Head and Neck Tumor and Lymph Segmentation in CT and PET Images

Xiao Sun⑩, Chengyang An⑩, and Lisheng Wang⁽✉⁾⑩

Department of Automation, Institute of Image Processing and Pattern Recognition,
Shanghai Jiao Tong University, Shanghai, People's Republic of China
lswang@sjtu.edu.cn

Abstract. Head and neck (H&N) cancer is one of the most prevalent cancers [1]. In its treatment and prognosis analysis, tumors and metastatic lymph nodes may play an important role but their manual segmentations are time-consuming and laborious. In this paper, we propose a coarse-to-fine ensembling framework to segment the H&N tumor and metastatic lymph nodes automatically from Positron Emission Tomography (PET) and Computed Tomography (CT) images. The framework consists of three steps. The first step is to locate the head region in CT images. The second step is a coarse segmentation, to locate the tumor and lymph region of interest (ROI) from the head region. The last step is a fine segmentation, to get the final precise predictions of tumors and metastatic lymph nodes, where we proposed a ensembling refinement model. This framework is evaluated quantitatively with aggregated Dice Similarity Coefficient (DSC) of 0.77782 in the task 1 of the HECKTOR 2022 challenge[2,3] as team SJTU426.

Keywords: Automatic segmentation · Head and neck cancer · Coarse-to-Fine

1 Introduction

In recent years, biomedical images play an important role in assisting diagnosis and treatment. The motivation is that biomedical images contain information that reflects underlying pathophysiology, and these relationships can be revealed by quantitative analysis [4]. For example, in the diagnosis and treatment of H&N cancer, locations of tumors and metastatic lymph nodes can be obtained from CT and PET images, which can help tailor treatment to specific patients. However, radiomics models often require the segmentation of the H&N tumor area in PET and CT images [5,6]. Manual segmentation is time-consuming and laborious. Therefore, an effective automatic segmentation framework for H&N cancer is needed.

Supported by Shanghai Jiao Tong University.

V. Andrearczyk et al. (Eds.): HECKTOR 2022, LNCS 13626, pp. 38–46, 2023.
https://doi.org/10.1007/978-3-031-27420-6_3

In the past few years, deep learning techniques based on convolutional neural networks(CNN) and Transformers [7] have achieved excellent results in many computer vision tasks for medical image analysis. In 2021, the HECKTOR (HEad and neCK TumOR segmentation and outcome prediction) challenge provided the opportunity of 3D segmentation of H&N tumors in CT and PET images, and a variety of deep learning methods have emerged [3]. Xie and Peng proposed a well-tuned patch-based 3D nnUNet [10] with Squeeze and Excitation (SE) normalization [11] and won the first place with a Dice Score of 0.7785 in the challenge [8]. Chenyang An proposed a coarse-to-fine framework with an ensemble of three U-Nets with SE normalization [9]. Of all the contestants, 3D U-Net and nnUNet are the most used frameworks, and SE normalization is also quite popular. However, in 2022, the HECKTOR challenge has some slight changes. The metastatic lymph nodes segmentation is added to the segmentation task, and no bounding boxes are provided.

In this paper, we propose a framework for precise segmentation of H&N primary tumor and metastatic lymph nodes by three progressive steps from the primitive CT and PET images. To achieve state of the art results, we combined the popular methods in 2021, the nnUNet and SE normalization, with nnFormer framework [12], an up-to-date technique of Transformers, and achieved Dice Similarity Coefficient of 0.77782 in the task 1 of the HECKTOR challenge 2022.

2   Method

We integrate the thought process of physician consultation into the model as a prior knowledge. More specifically, when a doctor reads a CT/PET image, he often finds the position of the head first, and then looks down and looks for the tumor in the neck region. Then for segmentation, we think it's necessary to apply a coarse-to-fine strategy, for the size of tumors and metastatic lymph nodes are relatively small to the overall CT/PET image, which might cause trouble on direct segmentation.

So, our method is a coarse-to-fine ensembling framework for H&N primary tumor and metastatic lymph nodes segmentation, which is divided into three stages: head locating, coarse segmentation and fine segmentation. Details will be described in Sect. 2.2.

2.1   Dataset

The challenge of HECKTOR2022 [3] provided 524 cases from 7 centers for training, and they does not contain any bounding box. Each case includes one 3D PET volume registered with a 3D CT volume, as well as a binary label volume with the annotated ground truth of the primary tumor and metastatic lymph nodes.

To maximize the learning of training dataset, we randomly selected 3 cases from each of the seven centers, 21 cases in total as the validation set and the rest 503 cases as the training set, with no testing set. It is noteworthy that

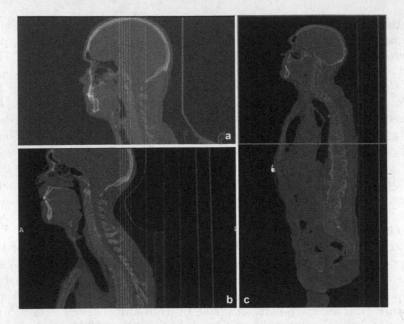

Fig. 1. Variation of data size from 7 centers. (a) only head region. (b) head and upper body. (c) whole upper body, and the resampled top region: red square. (Color figure online)

the training set is split into 5-fold cross-validation for any nnUNet or nnFormer framework.

2.2   Network Architecture

Head Locating. Since bounding box is not provided this year, and the data is from multiple centers, the size of the images varies, as shown in Fig. 1. The data sizes of the origin dataset mainly vary on the z axis. The shapes of images on the transverse plane are mostly 512 * 512, but the z axis varies from a few dozen to a few hundred. The difference in dimensions might cause trouble to the segmentation of neural network. Besides, although all heads are on the top of the image, but some heads are on the left, some are on the right (because they are from different centers). It would be a little hard to define a head size bbox automatically. So we decided to find our own bounding box for the head and neck region.

We decided to go from easy to complex, so the first step was to locate the region of the head. We do not need the precise head segmentation, all we need to know is the approximate position of the head. Here we apply a simple 3D U-Net with Residual connections [13] to locate the head region. Only CT images are used, which are resampled from the top of image with a spacing of (2, 2, 3) mm and a size of (256, 256, 128) with BSpline interpolation strategy, such as the red square in Fig. 1(c). 40 CT images were randomly selected from training set and we labeled the head region ourselves, which took just a few hours. All CT images were clipped (limited) into [−1000, 1000], and then normalized into [−1,1]. The detailed process is shown in Fig. 2.

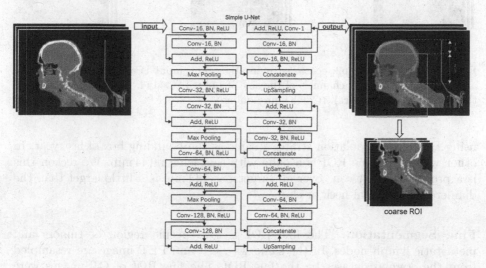

Fig. 2. Overview of step1: head locating. The input is CT patches of the size of $128 \times 256 \times 256$ All convolutional kernel sizes are $3 \times 3 \times 3$, and the number of convolution channels are depicted in each block. The blue square is calculation of the coarse ROI. (Color figure online)

After the head region is segmented, the bounding box of H&N region will be computed from the head label. To get a better view of H&N region, the point on the upper third of the vertical center axis of the head label is defined as the center of upper surface of the bounding box, as the blue square shown in Fig. 2. The bounding box is then resampled to (144, 144, 144) with a spacing of (2, 2, 2) mm using BSpline interpolation strategy, and it is defined as the coarse ROI. In other words, the coarse ROI is a cube with side length 288 mm, which is a little larger than the width of a normal head, so it should cover all the targets.

Coarse Segmentation. The coarse ROI is the region of the H&N where the tumors and metastatic lymph nodes belong. Here we apply the simple nnUNet as the coarse segmentation. CT and PET image are resampled from the original dataset by the coarse ROI. The coarse ROI of CT images were clipped into [−1000, 1000], and then normalized into [−1,1]. And the coarse ROI of PET images were normalized into [0,1], just at the coarse level.

After the coarse ROI is segmented, the fine ROI will be calculated from the tumor and lymph label. The center of all tumor and lymph labels is defined as the center of the fine ROI bounding box, as the blue square shown in Fig. 3. The bounding box is resampled to (144, 144, 144) with a spacing of (1, 1, 1) mm

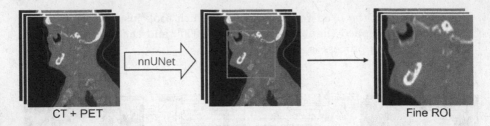

Fig. 3. Overview of step2: coarse segmentation. The input is a concatenation of CT and PET images as two channels of the size of $144 \times 144 \times 144$. And the calculation of the fine ROI (blue square). (Color figure online)

using BSpline interpolation strategy, same as the bounding box of last year. In other words, the fine ROI is a cube with side length 144 mm. We reckon this is a proper size that can cover all targets, as 144 mm is a little larget than the diameter of a normal neck or chin.

Fine Segmentation. The fine ROI is the specific region of tumors and metastatic lymph nodes. In the same way, CT and PET image are resampled from the original dataset by the fine ROI. The fine ROI of CT images were clipped into $[-1000, 1000]$, and then normalized into $[-1,1]$. And the fine ROI of PET images were normalized into $[0,1]$, just at the fine level.

In detail, fine segmentation consists of two parts, the distributed segmentation and the ensembling refinement (EF), as shown in Fig. 4.

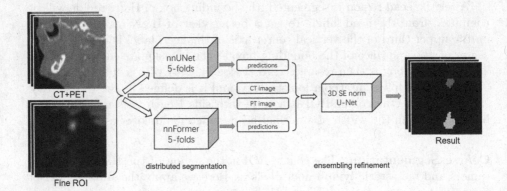

Fig. 4. Overview of step3: fine segmentation. The input is a concatenation of CT and PET images as two channels of the size of $144 \times 144 \times 144$.

The distributed segmentation consists of five-folds cross validation of both nnUNet and nnFormer, which means there are 10 models altogether. The input to the nnUNet and nnformer is a concatenation fine ROI of CT and PET images.

For each fold model, the output is a 2-channel binary label volume (the background challenge is abandoned), representing the segmentation of tumors and metastatic lymph nodes respectively. So, the output for all 10 models is 20 channels, and we concatenate them with original CT and PET images as the input to the EF model.

The EF model is a variant 3D U-Net with residual connections and SE normalization layers [11] with a input of 22 channels. The output of EF model is a three-channel binary label volume with a softmax activation function, representing background, tumors and metastatic lymph nodes. The network structure is shown in Fig. 5.

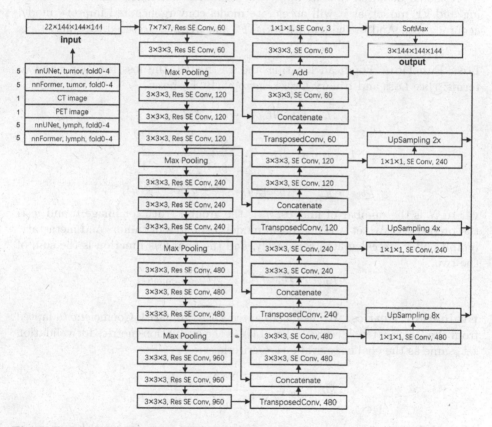

Fig. 5. Ensembling refinement (EF) architecture of fine segmentation. The input consists of both nnUnet and nnFormer 5-fold predictions and original CT and PET fine ROI images. The output consists of three channels representing background, tumors and metastatic lymph nodes respectively.

2.3   Training Details

All models were trained on two NVIDIA GeForce RTX 3090 (24 GB).

For head locating model, it was trained for 500 epochs using Adam optimizer, with a batch size of 1. The CosineDecayRestarts scheduler (from tensorflow2) is applied, with the initial learning rate of 10^{-4}, first-decay-steps of 50 , multiple times of 1.5 and decay rate of 0.96.

For nnUNet and nnFormer in coarse and fine segmentation, they were all trained on default settings, which means 1000 epochs for each fold with initial learning rate of 10^{-2} and Poly Learning Rate Decay.

For EF model, it was trained for 500 epochs using Adam optimizer, with a batch size of 1. The CosineAnnealingWarmRestarts scheduler (from pytorch) is applied, with initial learning rate of 10^{-4}, 30 iterations for the first restart.

It should be noted that we apply the cosine-restarts scheduler in head locating and EF model, as it will accelerate model convergence and improve model accuracy [16]. And in the team work, TF2 and pytorch are used together.

Loss Function. For head locating and EF model, the loss function has two terms: Dice Loss and Binary Cross Entropy (BCE) Loss:

$$L_{total} = \lambda L_{dice} + (1 - \lambda)L_{bce}$$

$$L_{dice}(y, \hat{y}) = 1 - \frac{2\sum_i^N y_i\hat{y}_i + 1}{\sum_i^N y_i + \sum_i^N \hat{y}_i + 1}$$

$$L_{bce}(y, \hat{y}) = -(y \log(\hat{y}) + (1 - y) \log(1 - \hat{y}))$$

where N is the number of images, y_i is the ground truth for image i, and $\hat{y}_i$ is the prediction. We set $\lambda = 0.5$ for all models. The loss of tumors and metastatic lymph nodes are calculated separately, and the final loss function is the sum of the two.

$$L_{final} = L_{total,tumor} + L_{total,lymph}$$

Evaluation Metrics. We use aggregated Dice Similarity Coefficient (adapted from the Aggregated Jaccard Index [14]) as the evaluation metrics for validation set, same as the challenge evaluation criteria.

$$DSC_{agg} = \frac{2\sum_i^N \sum_k \hat{y}_{i,k}y_{i,k}}{\sum_i^N \sum_k(\hat{y}_{i,k} + y_{i,k})}$$

where N is the number of images, $y_{i,k}$ is the ground truth for voxel k and image i, and $\hat{y}_{i,k}$ is the prediction.

Table 1. The validation set results of fine segmentation.

Agg. DSC	EF-60	EF-90	nnUNet	nnFormer
Tumor	0.8325	0.8282	0.8262	**0.8328**
Lymph	0.8324	**0.8335**	0.8325	0.8304

3 Results and Discussion

We tried two version of the EF model, EF-60 and EF-90. EF-60 is shown in Fig. 5 that the number of channels of the first convolution layer is 60. Similarly, the number of channels in the first convolution layer of EF 90 is 90, and the subsequent channels is doubled in the same way as EF-60.

The results of our validation set is summarized in Table 1. It can be concluded that nnUNet performs better in lymph segmentation and nnFormer performs better in tumor segmentation. These two models represent two kinds of segmentation methods, and we reckon they might have different priorities: Transformers in nnFormer may pay more attention to global information, as it can effectively capture and exploit long-term dependencies between pixels or voxels [12]. Convolutional layers in nnUNet may be better at local voxels, as the convolutions function on sliding windows across the image [15].

Our EF model is a balancing fusion of the two frameworks. For tumors, the EF model is a little behind nnFormer. For metastatic lymph nodes, EF-90 performs better than both nnUNet and nnFormer. However, in the final challenge submission, we still chose the predictions from EF model, because we hope that the EF model would generalize better to other dataset, as it performs on metastatic lymph nodes segmentation.

We combined tumor predictions from EF-60 and lymph predictions from EF-90 as our final submission, and we evaluated our predictions with agg DSC of 0.77782 in the task 1 of the HECKTOR 2022 Challenge. In addition, for comparisons with state-of-the-art methods, please refer to other participants of the challenge [3].

References

1. Parkin, D.M., Bray, F., Ferlay, J., Pisani, P.: Global cancer statistics. CA Cancer J. Clin. **55**(2), 74–108 (2005)
2. Oreiller, V., et al.: Head and neck tumor segmentation in PET/CT: the HECKTOR challenge. Med. Image Anal. **77**, 102336 (2022)
3. Andrearczyk, V., et al.: Overview of the HECKTOR challenge at MICCAI 2022: automatic head and neck tumor segmentation and outcome prediction in PET/CT. In: Head and Neck Tumor Segmentation and Outcome Prediction (2021). Springer, Heidelberg. https://doi.org/10.1007/978-3-030-98253-9_1
4. Gillies, R.J., Kinahan, P.E., Hricak, H.: Radiomics: images are more than pictures, they are data. Radiology **278**(2), 563–577 (2016)

5. Vallieres, M., et al.: Radiomics strategies for risk assessment of tumour failure in head-and-neck cancer. Sci. Rep. **7**(1), 1–14 (2017)
6. Bogowicz, M., et al.: Comparison of PET and CT radiomics for prediction of local tumor control in head and neck squamous cell carcinoma. Acta Oncologica **56**(11), 1531–1536 (2017)
7. Vaswani, A., et al.: Attention is all you need. Adv. Neural Inf. Process. Syst. **30** (2017)
8. Xie, J., Peng, Y.: The head and neck tumor segmentation based on 3D U-Net. In: 3D Head and Neck Tumor Segmentation in PET/CT Challenge, pp. 92–98. Springer, Cham (2021). https://doi.org/10.1007/978-3-030-98253-9_8
9. An, C., Chen, H., Wang, L.: A coarse-to-fine framework for head and neck tumor segmentation in CT and PET images. In: 3D Head and Neck Tumor Segmentation in PET/CT Challenge, pp. 50–57. Springer, Cham (2021). https://doi.org/10.1007/978-3-030-98253-9_3
10. Isensee, F., Jaeger, P.F., Kohl, S.A., Petersen, J., Maier-Hein, K.H.: nnU-Net: a self-configuring method for deep learning-based biomedical image segmentation. Nat. Methods **18**(2), 203–211 (2021)
11. Iantsen, A., Visvikis, D., Hatt, M.: Squeeze-and-excitation normalization for automated delineation of head and neck primary tumors in combined PET and CT images. In: Andrearczyk, V., Oreiller, V., Depeursinge, A. (eds.) HECKTOR 2020. LNCS, vol. 12603, pp. 37–43. Springer, Cham (2021). https://doi.org/10.1007/978-3-030-67194-5_4
12. Zhou, H.Y., Guo, J., Zhang, Y., Yu, L., Wang, L., Yu, Y.: nnformer: interleaved transformer for volumetric segmentation (2021). arXiv preprint arXiv:2109.03201
13. He, K., Zhang, X., Ren, S., Sun, J.: Deep residual learning for image recognition. In Proceedings of the IEEE Conference on Computer Vision and Pattern Recognition, pp. 770–778 (2016)
14. Kumar, N., Verma, R., Sharma, S., Bhargava, S., Vahadane, A., Sethi, A.: A dataset and a technique for generalized nuclear segmentation for computational pathology. IEEE Trans. Med. Imaging **36**(7), 1550–1560 (2017)
15. Krizhevsky, A., Sutskever, I., Hinton, G.E.: Imagenet classification with deep convolutional neural networks. Commun. ACM **60**(6), 84–90 (2017)
16. Loshchilov, I., Hutter, F.: SGDR: stochastic gradient descent with warm restarts. arXiv preprint arXiv:1608.03983 (2016)

A General Web-Based Platform for Automatic Delineation of Head and Neck Gross Tumor Volumes in PET/CT Images

Hao Jiang[1]([✉]), Jason Haimerl[1], Xuejun Gu[1,2], and Weiguo Lu[1,3]

[1] NeuralRad LLC, 8517 Exelsior Dr, Madison, WI 53717, USA
hao.jiang@neuralrad.com
[2] Stanford University, Stanford, CA 94305, USA
[3] University of Texas Southwest Medical Center, Dallas, TX 75235, USA

Abstract. Delineation of head and neck lesions are crucial for radiation treatment planning and follow-up studies. In this paper we developed an automated segmentation method for head and neck primary and nodal gross tumor volumes (GTVp and GTVn) segmentation in positron emission tomography/computed tomography (PET/CT) provided by the MICCAI 2022 Head and Neck Tumor Segmentation Challenge (HECKTOR 2022). Our segmentation algorithm takes nnU-Net as the backbone and uses dedicated pre- and post-processing to improve the auto-segmentation performance. The pipeline described achieved DSC results of 0.77212 (GTVp 0.77485 and GTVn 0.76938) in the testing dataset of HECTOR 2022. The developed auto-segmentation method is further extensively developed to a web-based platform to permit easy access and facilitate clinical workflow.

Keywords: Head and neck cancer · Segmentation · nnU-Net

1 Introduction

PET/CT is a widely used image modality for head and neck cancer (HNC) diagnosis and radiation treatment (RT) planning. With the strength of revealing metabolic and morphological tissue properties respectively, PET and CT modalities provide complementary and synergistic information for the segmentation of cancer lesions and tumor characteristics. In current clinical practice, tumor delineation is conducted manually, as human intelligence is required for processing the comprehensive information provided by radiological images, such as PET/CT and other clinical data such as age, gender, disease history. In the past few years, with the development of artificial intelligence (AI) technologies, particularly with the powerful modeling abilities of deep convolution neural network, deep-learning (DL) based auto-image segmentation has become a research upsurge [1–5].

In HNC RT field, DL-based auto-segmentation models have been developed and proven to be capable of delivering substantially better performance than conventional segmentation algorithms. Early developments used standard convolutional neural network classifiers with tailored pre- and postprocessing [6–9]. Later, the U-Net architecture

© The Author(s), under exclusive license to Springer Nature Switzerland AG 2023
V. Andrearczyk et al. (Eds.): HECKTOR 2022, LNCS 13626, pp. 47–53, 2023.
https://doi.org/10.1007/978-3-031-27420-6_4

[10] shown its promise in the medical image segmentation [11] and has been revised and refined for segmentation [12–19]. However, majority of these developments are focused on normal organ segmentation. Only a few tumor segmentation algorithms/tools have been proposed and developed [20–24], and the development is inadequate. The MICCAI Head and Neck Tumor segmentation challenge (HECKTOR) aims at promoting automatic AI models development for segmentation of Head and Neck (H&N) tumors in PET/CT images. In 2020, the challenge was focused on developing automatic bi-modal approaches for the segmentation of H&N primary Gross Tumor Volumes (GTVp), focusing on oropharyngeal cancers. In 2021, the scope of the competition was expanded by proposing the segmentation task on a larger population. For the 2022 challenge, the scope has been expanded even further by adding H&N nodal Gross Tumor Volumes (GTVn) segmentation. HECKTOR challenges provides a good platform for participants to compare a variety of approaches with the same data and evaluation criteria [2, 3].

This paper describes a customized deep-learning based segmentation pipeline developed based on nnU-Net with additional pre- and post-processing of input images and final output correspondingly. To further assisting clinical practice of the proposed model, we developed a general web-based platform to help the workflow of current PET/CT studies of data importing/exporting, segmentation, and visualization of the GTVs.

2   Materials and Methods

2.1   Network Architecture

The customized deep-learning based segmentation pipeline is built upon nnU-Net [1] with its backbone, the 3D full-res U-Net, as shown in Fig. 1.

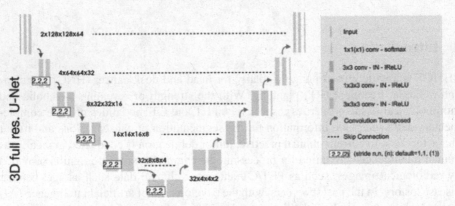

Fig. 1. nnU-Net 3D full resolution model architecture implemented in this study.

In this figure, the U-Net model comprises a down-sampling encoding path and up-sampling decoding path with skip connections in between, which helps to carry out feature extractions and retain the high-resolution segmentation results. Each steps of the encoding paths has corresponding convolutional block with the correct dimensions shown.

2.2 Data Preprocessing

2.2.1 Image Cropping

At the data preprocessing steps, for all 524 training cases, we first registered both CT and PET images and corresponding labels into the standard Head & Neck (H&N) geometry with ITK Mattes Mutual Information with Rigid body transformation. Then we calculated the median value of the centers of all GTVp region. Afterwards, the registered CT, PET and label images were cropped to a fixed $128 \times 128 \times 64$ ($1 \text{ mm} \times 1 \text{ mm} \times 3 \text{ mm}$) size using the median value of calculated center of GTVp region.

2.2.2 Normalization

Previous PET/CT GTV auto-segmentation studies have shown that PET signal plays a more important role than CT in the GTVp segmentation due to the high metabolic activities from tumor cells [4]. However due to the fact that test data come from multiple institutions and some of them differ from the training data, non-normalized PET data as input often results in less than ideal performance for segmentation results. In particular, this is often related with the fact that FDG intake for producing PET signal not only accumulates in tumors, but also appears in inflammation, biopsy, or benign regions.

Therefore, in our study, CT image values were normalized between 0 and 1 linearly whereas PET image values were treated differently in our submissions. In the first submission, we did not use any normalization method. For 2nd submissions, we used a clipping z-score normalization method. A lower bound and upper bound values were used to clip and average PET signals and the normalization was done using the median and standard deviations.

2.2.3 Outlier Removal

Finally, a separate script was carried out scanning all the cropped data to find out outliers and remove them from the training dataset, e.g., cases with GTVp volume less than 1% of the median value of all training GTVp volumes got removed. And cases with registration errors were removed in addition.

2.3 Training Procedure

The model was implemented in PyTorch and trained with 2500 epochs using SGD optimizers on a workstation with 7 NVIDIA GTX 1080 Ti (each having 11 GB GPU memory). The learning rate was reduced from the initial value of 1e−2 to 1e−3 during the training process. Several data augmentation methods implemented in nn-UNet such as scaling, rotation, gamma and mirror were utilized during data augmentation stage. The Dice loss is used as the loss function during the training process.

2.4 Post Processing Steps

During our 3rd submission, after model inference, several additional post processing steps were utilized to enhance the quality of the final results. We calculated the 0.5% of all the training data's GTVp (0.4cc) and GTVn (0.2cc) volumes and used them as the threshold to remove the isolated small-volume GTVp and GTVn labels, which could be potentially false-positive. Additionally, the distance between GTVn and GTVp were calculated, and we used 0.5% (12 cm) of such distance of the training data to remove the outlier cases. Finally, the final results were converted back the original un-cropped label using the header information of CT/PET images.

3   Dedicated Workflow Pipeline for Practical Clinical Usage

To facilitate the practical workflow in the clinics, we developed a general web-based framework which supports a workflow pipeline using our particular H&N model to quickly make predictions of GTVp and GTVn for the clinician. This web-based framework is an extension from our brain metastasis segmentation platform [25]. In detail, a DICOM (Digital Imaging and Communications in Medicine) server running on the workstation pulls in the patient images from hospital's PACS system or treatment planning system (TPS) using standard DICOM Send/DICOM Listener protocol. Then our proposed model will run predictions in the background and generate corresponding GTVp/GTVn DICOM-RT structures. A web-based WEBGL-implemented interface will display the overlay of registered CT/PET images, together with predicted GTVp/GTVn contours in both 3D and 2D (axial/sagittal/coronal views) visualizations. And the user can use this interface to quickly go through slides of the images to remove any false positives if present. Finally, the DICOM-RT structures with GTVp/GTVn will be automatically sent back to the TPS using the DICOM Send protocol. The corresponding web-based framework is shown in Fig. 2.

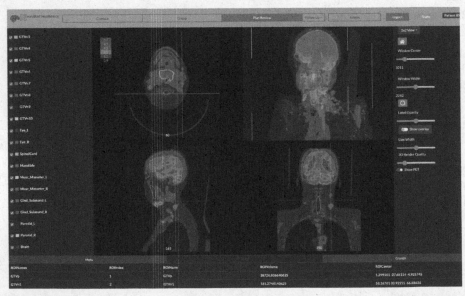

Fig. 2. Dedicated workflow module for the proposed H&N GTVp/GTVn segmentation model.

In addition, we also adopted and implemented our previously developed deep-learning based organ-at-risk (OAR) segmentation model [26] in this platform to assist the workflow during the treatment planning process.

4   Results and Discussion

Our three submitted model results are summarized in Table 1. Our first submission was using a model not using normalization of PET images but without any further post-processing. The second submission was based on the model using our clipping z-score normalization and without post-processing. The 3rd submission was the model with normalization and post-processing steps described in Sect. 2.

Table 1. Comparison of Dice results between model trained using the original PET data and normalized PET data

	DSC (GTVp)	DSC (GTVn)	DSC (Average)
Model trained/un-normalized PET data (1st submission)	0.74618	0.74507	0.74563
Model trained/normalized PET data (2nd submission)	0.76555	0.75723	0.76139
With post processing (3rd submission)	0.77485	0.76938	0.77212

Our 1st submission not using normalization achieved average Dice at 0.74563. At 2nd submission, the model trained using our clipping z-score normalized PET input achieved better average Dice at 0.76139, showing improvements using normalization. The best outcome is from our 3rd submission trained with the normalized PET images model and post-processing results. The achieved average Dice at 0.77212 showed the significance of using additional post-processing steps for improvement of the results.

The intensity value of FDG-PET image as used in this study represents the FDG concentration that reaches the lesion, which is highly varied and patient-dependent. Such patient-dependent variation is due to the combination of different physiological factors, such as the blood glucose level, medications, age, gender and etc. [27]. On the other hand, the convolution base nnU-Net heavily relies on image voxels' intensity values to extract image features and consequently determine the class that each voxel belongs to. For a robust model training, image intensity normalization is a necessity to help to minimize the impact of intensity variation.

The post-processing steps adopted in this project is derived from domain knowledge. As nnU-Net generates the label using the pixel-by-pixel classification results, which potentially includes small-volume false positives. While in clinic, isolated small-volume lesions are rare. Similarly, clinically, nodal involvement often happens in the nearby lymph nodes rather than distance-away. Thus excluding those isolated small volumes and distant GTVn aligns with evidence-based clinical practice and could better match with manually-delineated ground truth contours. In total, we removed 4 nodes with distance to GTVp >120 mm and 2 nodes with its volume <200 mm^3. Out of total of 693 nodes detected.

5 Conclusions

An automatic workflow pipeline and a general web-based framework were developed to assist clinicians to quickly contour H&N GTVp and GTVn using a customized segmentation pipeline developed based on nnU-Net. The Dice coefficients (DSC) reached 0.77485 and 0.76938 for GTVp and GTVn respectively in the training dataset.

References

1. Xing, L., Giger, M.L., Min, J.K. (eds.): Artificial Intelligence in Medicine: Technical Basis and Clinical Applications. Elsevier Science, St. Louis (2020)
2. Krizhevsky, A., Sutskever, I., Hinton, G.E.: Imagenet classification with deep convolutional neural networks. Paper Presented at: Advances in Neural Information Processing Systems (2012)
3. Russakovsky, O., et al.: ImageNet large scale visual recognition challenge. Int. J. Comput. Vis. **115**(3), 211–252 (2015). https://doi.org/10.1007/s11263-015-0816-y
4. Long, J., Shelhamer, E., Darrell, T.: Fully convolutional networks for semantic segmentation. Paper Presented at: Proceedings of the IEEE Conference on Computer Vision and Pattern Recognition (2015)
5. Ronneberger, O., Fischer, P., Brox, T.: U-Net: convolutional networks for biomedical image segmentation. In: Navab, N., Hornegger, J., Wells, W., Frangi, A. (eds.) MICCAI 2015. LNCS, vol. 9351, pp. 234–241. Springer, Cham (2015). https://doi.org/10.1007/978-3-319-24574-4_28
6. Močnik, D., Ibragimov, B., Xing, L., et al.: Segmentation of parotid glands from registered CT and MR images. Phys. Med. **52**, 33–41 (2018)
7. Ren, X., Xiang, L., Nie, D., et al.: Interleaved 3D-CNNs for joint segmentation of small-volume structures in head and neck CT images. Med. Phys. **45**(5), 2063–2075 (2018)
8. Ibragimov, B., Xing, L.: Segmentation of organs-at-risks in head and neck CT images using convolutional neural networks. Med. Phys. **44**(2), 547–557 (2017)
9. Zhong, T., Huang, X., Tang, F., Liang, S., Deng, X., Zhang, Y.: Boosting-based cascaded convolutional neural networks for the segmentation of CT organs-at-risk in nasopharyngeal carcinoma. Med. Phys. **46**, 5602–5611 (2019)
10. Ronneberger, O., Fischer, P., Brox, T.: U-Net: convolutional networks for biomedical image segmentation. arXiv e-prints (2015). https://ui.adsabs.harvard.edu/abs/2015arXiv150504597R. Accessed 01 May 2015
11. De Fauw, J., Ledsam, J.R., Romera-Paredes, B., et al.: Clinically applicable deep learning for diagnosis and referral in retinal disease. Nat. Med. **24**(9), 1342–1350 (2018)
12. Tong, N., Gou, S., Yang, S., Ruan, D., Sheng, K.: Fully automatic multi-organ segmentation for head and neck cancer radiotherapy using shape representation model constrained fully convolutional neural networks. Med. Phys. **45**(10), 4558–4567 (2018)
13. Liang, S., et al.: Deep-learning-based detection and segmentation of organs at risk in nasopharyngeal carcinoma computed tomographic images for radiotherapy planning. Eur. Radiol. **29**(4), 1961–1967 (2018)
14. Wang, Y., Zhao, L., Wang, M., Song, Z.: Organ at risk segmentation in head and neck CT images using a two-stage segmentation framework based on 3D U-Net. IEEE Access. **7**, 144591–144602 (2019)

15. Men, K., Geng, H., Cheng, C., et al.: Technical note: more accurate and efficient segmentation of organs-at-risk in radiotherapy with convolutional neural networks cascades. Med. Phys. **46**(1), 286–292 (2019)
16. Tappeiner, E., et al.: Multi-organ segmentation of the head and neck area: an efficient hierarchical neural networks approach. Int. J. Comput. Assist. Radiol. Surg. **14**(5), 745–754 (2019)
17. Rhee, D.J., Cardenas, C.E., Elhalawani, H., et al.: Automatic detection of contouring errors using convolutional neural networks. Med. Phys. **46**(11), 5086–5097 (2019)
18. Tang, H., Chen, X., Liu, Y., et al.: Clinically applicable deep learning framework for organs at risk delineation in CT images. Nat. Mach. Intell. **1**(10), 480–491 (2019)
19. van Rooij, W., Dahele, M., Ribeiro Brandao, H., Delaney, A.R., Slotman, B.J., Verbakel, W.F.: Deep learning-based delineation of head and neck organs at risk: geometric and dosimetric evaluation. Int. J. Radiat. Oncol. Biol. Phys. **104**(3), 677–684 (2019)
20. Guo, Z., Guo, N., Gong, K., Zhong, S., Li, Q.: Gross tumor volume segmentation for head and neck cancer radiotherapy using deep dense multi-modality network. Phys. Med. Biol. **64**(20), 205015 (2019)
21. Lin, L., Dou, Q., Jin, Y.M., et al.: Deep learning for automated contouring of primary tumor volumes by MRI for nasopharyngeal carcinoma. Radiology **291**(3), 677–686 (2019)
22. Men, K., Chen, X., Zhang, Y., et al.: Deep deconvolutional neural network for target segmentation of nasopharyngeal cancer in planning computed tomography images. Front Oncol. **7**, 315 (2017)
23. Jin, D., Guo, D., Ho, T.-Y., et al.: DeepTarget: Gross tumor and clinical target volume segmentation in esophageal cancer radiotherapy. Med. Image Anal. **68**, 101909 (2021)
24. Cardenas, C.E., Beadle, B.M., Garden, A.S., et al.: Generating high-quality lymph node clinical target volumes for head and neck cancer radiation therapy using a fully automated deep learning-based approach. Int. J. Radiat. Oncol. Biol. Phys. **109**(3), 801–812 (2021)
25. Yang, Z., Liu, H., Liu, Y., et al.: A web-based brain metastases segmentation and labeling platform for stereotactic radiosurgery. Med. Phys. **47**(8), 3263–3276 (2020)
26. Chen, H., Lu, W., Chen, M., et al.: A recursive ensemble organ segmentation (REOS) framework: application in brain radiotherapy. Phys. Med. Biol. **64**(2), 025015 (2019)
27. Sprinz, C., Zanon, M., Altmayer, S., et al.: Effects of blood glucose level on 18F fluorodeoxyglucose (18F-FDG) uptake for PET/CT in normal organs: an analysis on 5623 patients. Sci. Rep. **8**(1), 2126 (2018)
28. Isensee, F., Jaeger, P.F., Kohl, S.A., Petersen, J., Maier-Hein, K.H.: nnU-Net: a self-configuring method for deep learning-based biomedical image segmentation. Nat. Methods **18**, 203–211 (2021)
29. Andrearczyk, V., et al.: Overview of the HECKTOR challenge at MICCAI 2022: automatic head and neck tumor segmentation and outcome prediction in PET/CT. In: Andrearczyk, V., Oreiller, V., Hatt, M., Depeursinge, A. (eds.) HECKTOR 2022. LNCS, vol. 13626, pp. 1–30. Springer, Cham (2023). https://doi.org/10.1007/978-3-031-27420-6_4
30. Oreiller, V., et al.: Head and neck tumor segmentation in PET/CT: the HECKTOR challenge. Med. Image Anal. **77**, 102336 (2022)

Octree Boundary Transfiner: Efficient Transformers for Tumor Segmentation Refinement

Anthony Wang[✉], Ti Bai, Dan Nguyen, and Steve Jiang

UT Southwestern Medical Center, Dallas, TX 75390, USA
anthonyindeepspace@gmail.com

Abstract. In this paper, we create a fully autonomous system that segments primary head and neck tumors as well as lymph node tumors given only FDG-PET and CT scans without contrast enhancers. Given only these two modalities, the typical Dice score for the state-of-the-art (SOTA) models lies below 0.8, below what it would be when including other modalities due to the low resolution of PET scans and noisy non-enhanced CT images. Thus, we seek to improve tumor segmentation accuracy while working with the limitation of only having these two modalities. We introduce the Transfiner, a novel octree-based refinement system to harness the fidelity of transformers while keeping computation and memory costs low for fast inferencing. The observation behind our method is that segmentation errors almost always occur at the edges of a mask for predictions from a well-trained model. The Transfiner utilizes base network feature maps in addition to the raw modalities as input and selects regions of interest from these. These are then processed with a transformer network and decoded with a CNN. We evaluated our framework with Dice Similarity Coefficient (DSC) 0.76426 for the first task of the Head and Neck Tumor Segmentation Challenge (HECKTOR) and ranked 6th.

Keywords: Transformer · Octree · U-Net

1 Introduction

Manual tumor segmentation is commonly used for radiotherapy treatment planning, but it is a time-consuming, subjective (resulting in low inter-observer agreement), and labor-intensive task. This is a problem because variations in tumor segmentation can lead to differences in the prescription of radiation doses, killing healthy tissue, leaving tumor tissue untouched, and ultimately impacting a patient's outcome. Automated tumor segmentation can reduce the time required for treatment planning and improves the consistency of results.

There has been recent interest in using deep learning methods for automated tumor segmentation. However, these methods often require multiple types of imaging modalities, such as PET, CT, and many variations of MRI [8]. This can

V. Andrearczyk et al. (Eds.): HECKTOR 2022, LNCS 13626, pp. 54–60, 2023.
https://doi.org/10.1007/978-3-031-27420-6_5

be problematic, as obtaining all of these types of images for every patient can be time-consuming and expensive. This challenge provides only FDG-PET and non-enhanced CT, making it harder for accurate segmentation, particularly on the edges due to noisy scans.

A key empirical observation we make is that segmentation error almost always occurs at the edges of a mask for predictions from a well-trained model, rather than completely missing tumor tissue. Our proposal aims to fix this.

1.1 Related Work

In [4], the authors introduce a transfiner which is similar to our proposed method. Both methods are designed to improve the quality of segmentation masks, with a focus on the edges of the masks. The authors use a quadtree decomposition to represent the incoherence maps, which is similar to our octree decomposition. We use the output segmentation masks to generate incoherence maps, while the authors use convolutional layers to predict where the incoherent regions are. We also use U-Net feature maps as input to the Transfiner, whereas the authors use pyramid feature maps.

In [5], the authors introduce the Vision Transformer (ViT), which creates transformer input tokens by patchifying an image into a sequence of non-overlapping patches, and applying a linear layer to each flattened patch. These tokens are then constructed using a standard transformer architecture. We adapt this technique of patches in our model, but we use a different scheme for selecting patches, as well as a different method for turning these patches into tokens.

2 Dataset

The training dataset contains 524 patients from 7 hospitals. The data originates from FDG-PET and low-dose non-contrast-enhanced CT images (acquired with combined PET/CT scanners) of the H&N region. The segmentation maps contains 3 classes - primary tumor (GTVt), tumor-involved lymph nodes (GTVn), and background. The testing dataset contains 356 patients from 3 hospitals. One notable aspect of the data is that the test dataset and training dataset both contain patients from the "MDA" center. The test dataset contains 200 patients from the center while the training dataset contains 201 patients from there. [2]

2.1 Data Preprocessing

We employ a variety of preprocessing operations designed to normalize inter-subject variation and to enhance relations between co-registered PET and CT images. These include isotropic linear resampling to 1 mm cubic voxels, modality-wise clipping of input to the 0.5th and 99.5th percentile, and modality-wise Z-score normalization of input.

2.2 Data Augmentation

For data augmentation for both models, we scale, rotate and mirror the 3D volumetric images. We choose small scale and angle factors to augment small variations in the original data distributions. We also use gamma correction to augment the contrast of image intensities (Table 1).

Table 1. Data augmentation details

Operation	Details
Rotation	Rotation about x, y and z axis by a random angle in (−30, 30) degrees
Scaling	Scaling by a random factor between 0.7 and 1.4
Gamma	Gamma correction with a random factor between 0.7 and 1.4
Mirroring	Mirroring about x, y and z axis with a probability of 0.5

3 Methods

We use a base nnU-Net [1] to extract feature maps from the input image. We then pass these through a Transfiner, which uses a Laplacian pyramid to generate regions of interest. Our Transfiner reads input similar to the patchifying operation in a vision transformer, but uses an octree decomposition of multiple layers of interest to select relevant patches instead of patchifying the whole image. This allows us to select both large and small patches which are relevant to the task at hand, while conserving memory and computational resources. A 3-d cosine positional embedding is added to the Transfiner input, which encodes the position of each patch in 3-dimensional space. The Transfiner is then passed through a transformer network, which attends to relevant patches and produces an output representation. This output is then decoded by a CNN, which outputs the final segmentation mask (Fig. 1).

3.1 Losses

We use a combination of Dice and cross entropy (CE) losses to train our model.
The Dice loss is defined as:

$$\mathcal{L}_{dice} = 1 - \frac{2\sum_i p_i g_i}{\sum_i p_i + \sum_i g_i},$$

where p is the predicted mask and g is the ground truth mask.
The cross entropy loss is defined as:

$$\mathcal{L}_{ce} = \sum_i -g_i log(p_i),$$

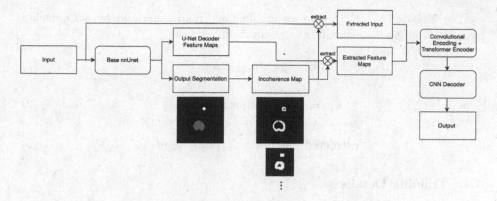

Fig. 1. Diagram showing complete architecture

Our loss is an evenly weighted sum of the Dice and cross entropy losses:

$$\mathcal{L}_{total} = \alpha\mathcal{L}_{ce} + (1 - \alpha)\mathcal{L}_{dice},$$

where $\alpha = 0.5$.

3.2   Incoherence Maps

To obtain a Laplacian Pyramid [7] from our initial predictions, we start by generating a Gaussian Pyramid, consisting of N layers of iteratively downsampled images (only downsampled in spatial dimensions), starting with the original image. Given our downsampling operator, D, we can express the Gaussian Pyramid as:

$$G = \{X, D(X), D^2(X), ..., D^{N-1}(X)\}$$

The Laplacian Pyramid is then generated by subtracting each layer of the Gaussian Pyramid from the next layer up, and then upsampling, U, the result. This can be expressed as:

$$L = \{X - U(D(X)), D(X) - U(D^2(X)), ..., D^{N-2}(X) - U(D^{N-1}(X))\}$$

Finally, the incoherence map is obtained through a simple thresholding operation of the Laplacian Pyramid to limit the size of the incoherence map.

3.3   Patch Selection

The following describes our patch selection algorithm. Let $X_1, X_2 \in \mathbb{R}^{H \times W \times D \times C_{in}}$ be an intermediate segmentation map and a feature map from the base network.

Let $Y \in \mathbb{R}^{H \times W \times D \times C_{in}}$ be an incoherence map generated by a Laplacian pyramid from the intermediate segmentation.

$$idx = \text{pick}(\mathbf{a} \neq 0, \mathbf{a} \in Y) \tag{1}$$

$$idx = \text{shuffle}(\mathbf{e})[: \text{max patches}] \tag{2}$$

$$\text{extracted patches} = X_1[idx], X_2[idx] \tag{3}$$

3.4   Training Details

In our testing, we used V100 GPUs for training. We use an SGD optimizer with momentum of 0.99. The training duration is fixed at 1000 epochs and during training we employ One cycle learning rate. The initial and final learning rates are 0.01 and 0.00005 respectively. The learning rates go through a peak during the 100th epoch at 0.015. We use a batch size of 2 and equal weights for the Dice and cross entropy losses. We train a patch-based model with patch size of 128 and patch overlap of 64. We choose a 5 model logit-averaging ensemble to improve prediction accuracy (Table 2).

Table 2. Training details

Feature	Detail
Optimizer	SGD + momentum
Learning rate scheduler	1 cycle lr [6]
Initial learning rate	0.01
Peak learning rate	0.015 (epoch 100)
Final learning rate	0.00005
Training duration	1000 epochs
Batch size	2
Loss weights	Dice + CE loss
Patch size	128
Patch overlap	64
Models to average	5

4   Results

We found that there was only a minor difference in results between using nnU-Net and Transfiner, although more significant differences were found in the primary tumor Dice scores in our cross-validation testing. One interesting statistic to note is that our Transfiner model improved primary tumor loss in our cross-validation testing and leads Dice score in primary segmentation, but primary Dice is lower on official testing data.

This supports our hypothesis that the Transfiner refines edges, although it also suggests that it may be prone to overfitting. In the future, we would like to investigate qualitatively on our model's performance on the testing data, and inspect edge-refining performance on data from unseen testing centers.

With the Transfiner, we were able to achieve Dice scores of 0.76426 for the first task of HECKTOR, ranking us 6th place. We noticed that our primary tumor DSC was much lower than our lymph node tumor Dice score, which was very unusual compared to the rest of the top teams. In terms of lymph-node involved DSC (gTVn), our model ranked 3rd place. Our 5-fold cross validation results are listed in the table below (Table 3).

Table 3. 5 fold cross validation results.

Model	Details	gTVp	gTVn
nnU-Net	0.7723	0.7748	0.7697
Transfiner	0.7794	0.7812	0.7652

5   Conclusion

In this paper, we have tested the Transfiner, a octree-based refinement system designed to harness the fidelity of transformers while keeping computation and memory costs low for fast inference. We compared the Transfiner to a base nnU-Net model and found that the Transfiner was able to improve segmentation accuracy. We hypothesize that the success of this network stems from the use of transformer networks, which are able to capture global dependencies in the data. We plan to further investigate the Transfiner by testing it on other datasets and tasks. Additionally, we plan to investigate the effect of different transformer sizes and architectures on the Transfiner's performance in comparison to changing the size of the base network. We believe that the Transfiner has potential to be a powerful tool for segmentation and we hope to continue exploring its capabilities in future work.

References

1. Isensee, F., Jaeger, P.F., Kohl, S.A., Petersen, J., Maier-Hein, K.H.: nnU-Net: a self-configuring method for deep learning-based biomedical image segmentation. Nat. Methods **18**, 1–9 (2020)

2. Andrearczyk, V., et al.: Overview of the HECKTOR challenge at MICCAI 2022: automatic head and neck tumor segmentation and outcome prediction in PET/CT. In: Head and Neck Tumor Segmentation and Outcome Prediction. Springer, Heidelberg (2023). DOI: https://doi.org/10.1007/978-3-030-98253-9_1

3. Oreiller, V., et al.: Head and neck tumor segmentation in PET/CT: the HECKTOR challenge. Med. Image Anal. **77**, 102336 (2022)

4. Ke, L., Danelljan, M., Li, X., Tai, Y., Tang, C., Yu, F.: Mask transfiner for high-quality instance segmentation (2021). arXiv https://arxiv.org/abs/2111.13673

5. Dosovitskiy, A., et al.: An image is worth 16×16 words: transformers for image recognition at scale. CoRR. abs/2010.11929 (2020). https://arxiv.org/abs/2010.11929

6. Smith, L., Topin, N.: Super-convergence: very fast training of neural networks using large learning rates (2017). arXiv, https://arxiv.org/abs/1708.07120

7. Burt, P., Adelson, E.: The Laplacian pyramid as a compact image code. IEEE Trans. Commun. **31**, 532–540 (1983)

8. Faraji, F., Gaba, R.: Radiologic modalities and response assessment schemes for clinical and preclinical oncology imaging. Front. Oncol. **9** (2019), https://doi.org/10.3389/fonc.2019.00471

Head and Neck Primary Tumor and Lymph Node Auto-segmentation for PET/CT Scans

Arnav Jain, Julia Huang, Yashwanth Ravipati, Gregory Cain, Aidan Boyd, Zezhong Ye, and Benjamin H. Kann[✉]

Artificial Intelligence in Medicine Program, Mass General Brigham, Harvard Medical School, Boston, MA 02115, USA
{ajain19,jhuang58,yravipati,aboyd7,zye2}@bwh.harvard.edu,
gbcain@mgh.harvard.edu, benjamin_kann@dfci.harvard.edu
http://www.aim.hms.harvard.edu/

Abstract. Segmentation of head and neck (H&N) cancer primary tumor and lymph nodes on medical imaging is a routine part of radiation treatment planning for patients and may lead to improved response assessment and quantitative imaging analysis. Manual segmentation is a difficult and time-intensive task, requiring specialist knowledge. In the area of computer vision, deep learning-based architectures have achieved state-of-the-art (SOTA) performances for many downstream tasks, including medical image segmentation. Deep learning-based auto-segmentation tools may improve efficiency and robustness of H&N cancer segmentation. For the purpose of encouraging high performing methods for lesion segmentation while utilizing the bi-modal information of PET and CT images, the HEad and neCK TumOR (HECKTOR) challenge is offered annually. In this paper, we preprocess PET/CT images and train and evaluate several deep learning frameworks, including 3D U-Net, MNet, Swin Transformer, and nnU-Net (both 2D and 3D), to segment CT and PET images of primary tumors (GTVp) and cancerous lymph nodes (GTVn) automatically. Our investigations led us to three promising models for submission. Via 5-fold cross validation with ensembling and testing on a blinded hold-out set, we received an average of 0.77 and 0.70 using the aggregated Dice Similarity Coefficient (DSC) metric for primary and node, respectively, for task 1 of the HECKTOR2022 challenge. Herein, we describe in detail the methodology and results for our top three performing models that were submitted to the challenge. Our investigations demonstrate the versatility and robustness of such deep learning models on automatic tumor segmentation to improve H&N cancer treatment. Our full implementation based on the PyTorch framework and the trained models are available at https://github.com/xmuyzz/HECKTOR2022 (Team name: AIMERS).

Keywords: Segmentation · HECKTOR · Deep learning

1 Introduction

Head and Neck cancer is one of the most prevalent cancers in the world, with over 630,000 new cases annually [7]. Computed tomography (CT) and positron emission tomography (PET) scans are currently used to diagnose and propose treatment plans. However, current methods of diagnosis and treatment, such as radiotherapy, require manual annotations of tumors from experienced oncologists. Deep learning approaches have shown great promise in automatically segmenting medical data, allowing for a more efficient radiation therapy workflow to exist.

The Medical Image Computing and Computer Assisted Intervention Society (MICCAI) created the HEad and NeCK TumOR(HECKTOR) Segmentation Challenge to explore pathways for autosegmentation of both primary tumors and metastatic lymph nodes [1,6]. It utilizes high quality, multi-institutional data containing both PET and CT scans. In this paper, we explore using state of the art segmentation models inserted into self-configuring networks to achieve a high segmentation accuracy.

2 Materials and Methods

2.1 Dataset

The HECKTOR 2022 Challenge training dataset is comprised of 523 PET and CT scans gathered from multiple institutes. For each scan, there is an associated mask, containing masks of both the primary tumor and/or lymph nodes.

Upon downloading the dataset, we manually split the dataset into a training set of 453 (87%) and a hold-out set of 70 (13%) scans, stratifying by contributing institution.

2.2 Preprocessing

Prior to feeding the data into our models for training and testing, we resampled both the image and label NIFTI files to $1 \times 1 \times 3$ spacing using SimpleITK linear interpolation. This was done to ensure better generalization and a more manageable spacing size. We then cropped scans that covered the entire body (as opposed to only head/neck), as indicated by the z-axis dimension, to 65% of their original slice length, to remove superfluous information from inferior anatomic regions like the abdomen and pelvis. A fixed reference file was selected and we performed rigid registration for all data files according to this template, using previously described methods [8]. We chose CHUM-021 as our reference image from which all other data files were rigidly registered, and coordinate correspondence was achieved between all the scans. After registration, the files were cropped to either $172 \times 172 \times 76$ or $160 \times 160 \times 64$ resolution using a custom bounding box procedure following identification of the center of the skull. We used OpenCV to identify the largest elliptical contour in the top two-thirds of the image by comparing the area: perimeter ratios for all contours. We then performed a centroid calculation of the skull ellipse to center the crop around the head region of each scan. The CT scans are then clipped and normalized based on the mean, median, and upper and lower percentiles as done in the nnU-Net methodology. All preprocessing steps are synchronously done with the PET scans, apart from the normalization step, to ensure that the PET and CT scans are aligned and maintain the same dimensions (Fig. 1).

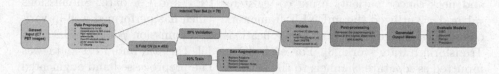

Fig. 1. Overall pipeline of our process, from dataset input to processing to model training to evaluation. (1): The HECKTOR dataset is comprised of PET and CT scans. (2): We perform a series of preprocessing steps including: (a) Resampling the scans to 1×1×3, (b) Cropping scans to 65% on its z-axis from the top, (c) Using rigid registration to register all scans to a reference scan, (d) Using eliptical contouring to center the scan around the patients head, (e) Using CT clipping on the CT scans to help normalize the data (Note: CT and PET scans undergo the same resampling, registration, and resampling steps, but PET doesn't undergo the clipping and normalization steps that the CT scans go through). (3) We then split off the data into 70 scans for the internal testing set and 353 scans which is used for 5 fold cross validation for the models. (4): The models are then trained using nnU-Net, nnM-Net, and Swin UNETR architectures, as described in Sects. 2.5–2.7. (5): The outputs then undergo post-processing, which performs the inverse of the preprocessing steps and reverts the scans back to its original spacing and dimensions. (6): The final output masks are generated. (7): These masks then undergo evaluation on the testing set to output metrics including Dice Coefficients, Jaccard Index, Recall, and Precision for each model tested.

For all datasets, preprocessed files were stored into Google Drive folders named imagesTs, imagesTr, labelsTs, and labelsTr for test images, training images, test labels, and training labels, respectively. Preprocessing was run on both Google Colab as well as Linux machines equipped with Nvidia RTX Titan GPUs.

2.3 Augmentations

The CT and PET scans are converted into 3D NumPy arrays, on which augmentations are done. We used default augmentations on the nnU-Net models [4] (model details below), consisting of modifications such as rotations of the x, y, and/or z axes, scaling, gaussian noise, gaussian blur, random gamma, random flipping, and elastic transformations. Furthermore, the nnU-Net implements CT clipping and PET normalization in the preprocessing stage.

On the models trained without the nnU-Net, we used random cropping to a region of $160 \times 160 \times 64$, random Gaussian noise, random rotation of up to 8° on all three axis, and a random elastic transform. Although we experimented with CT thresholding and PET normalization, we found no significant increase/slight decrease in Dice score.

2.4   Architecture Overview

We used three deep learning frameworks, all based on an encoder-decoder style architecture for autosegmentation of primary tumors and lymph nodes of head and neck cancer patients using co-registered PET and CT data. Submissions #1 and #2 utilized the nnUNET framework, which has performed well in various medical image segmentation applications [4]. Submission #3 utilized a Swin Transformer framework [3]. For each 3D model, the CT and PET data was inputted as separate channels to the network. The preprocessed and augmented PET and CT scan for each patient is concatenated along a new axis to create a 4D array of dimensions $172 \times 172 \times 76 \times 2$, where the fourth axis represents the two channels of PET and CT data, which will serve as the input dimensions for the model.

Model submissions #1 (nnU-Net 3D) and #2 (nnMNet) were trained using a combined Cross Entropy and Dice Loss function. The Swin Transformer model (submission #3) was trained using Unified Focal Loss [9]. The models were trained with 362 training and 91 validation cases, and submissions #1 and #2 utilized 5-fold cross validation on the combined training/validation set, after which the five models were ensembled to undergo testing on the internal testing set.

2.5   nnU-Net

Here we describe the specifics of the nnU-Net [4], a semi-automated UNet framework developed by Isensee et al. specifically designed for medical images to adapt to a large variety of modalities. The nnU-Net is known to be the first technique developed to condense and automate decisions for segmentation on biomedical datasets. nnU-Net was tested on 23 public biomedical segmentation datasets and outperforms most existing methods. In addition, this architecture is reproducible and practical to use, as codes along with training and evaluation procedures are published on GitHub. Lastly, nnU-Net's flexible framework and structure allows for easy integration with other architectures such as Transformer or the MNet [2], which we used in our research.

The nnU-Net, a type of convolutional neural network architecture built on the PyTorch framework, automatically configures the preprocessing, architecture parameters, training, and post-processing steps for any task, covering the entire segmentation pipeline that matches each dataset's unique properties. Firstly, nnU-Net extracts a dataset fingerprint, which includes parameters specific to each dataset such as image size and intensity information to create different U-Net configurations; 2D, 3D full-res for full resolution images, and 3D cascade [3]. For initial training and testing, we experiment with both the 2D and 3D full-res models; however, for final submission to the HECKTOR challenge, we only utilize 3D. The nnU-Net places domain knowledge into three parameter

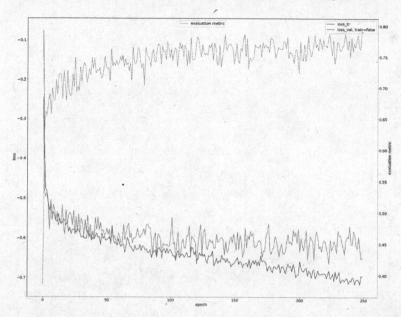

Fig. 2. Training and validation losses and Dice plots by epoch for fold 0 of the nnUNet. It runs until epoch 250.

groups, which serves to set dataset properties such as batch size and patch size during method design (Fig. 2).

We trained both the 2D and 3D full-resolution nnU-Nets for 200–300 epochs for each fold. We used default batch sizes of nnU-Net: 2 for 3D and 50 for 2D. The initial learning rate was 0.01 and decays using polynomial decay. We used stochastic gradient descent (SGD) with Nesterov momentum ($\mu = 0.99$) as the optimizer. In addition, we used a combination of cross-entropy and Dice loss as the loss function. For the environment, we utilized GPUs from Google Colab Pro Plus and Linux for running the models. We used Python 3.7.13 and PyTorch version 1.5.1 with CUDA 10.1.

We tested the models on a internal holdout testing set, evaluating each model based on the aggregated Dice score for primary tumor and lymph node.

2.6 MNet

In addition, we investigated the MNet [2] architecture for submission #2, developed by Dong et al. MNet fuses 2D and 3D data features to balance out inter- and intra-slice representation, as it contains multi-dimensional convolutions that make the selection of different representations more flexible. MNet has shown superior results to other segmentation architectures such as the UNet on other datasets [2]. In addition, because MNet combines 2D and 3D features, it utilizes both their advantages: in 2D views, regions are easily recognized and thus lead to high segmentation accuracies, and in 3D views, the organ contours have high smoothness (Fig. 3).

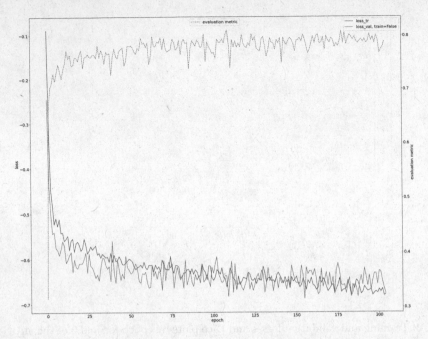

Fig. 3. Training and validation losses and Dice plots by epoch for fold 1 of the nnMNet. It runs until epoch 200.

In our work, we integrated the MNet into the nnU-Net architecture, coined nnM-Net, replacing the internal network with the MNet. This approach combines the self-configuration advantages of the nnU-Net with the more robust architecture of the MNet. We utilized the same default nnU-Net augmentations, batch sizes, parameters, and the same training procedure and environment mentioned in the nnU-Net section.

2.7   Swin Transformer

Lastly, we experimented with utilizing a Swin Transformer UNet [3] as our segmentation architecture. Swin Transformer models have shown great success in image classification, achieving superior accuracies compared to traditional convolutional networks on ImageNet. We experimented with the SwinUNET architecture developed by MONAI, which utilizes Swin transformers in the encoder and FCNN layers in the decoder. It has shown promising results on other medical datasets, such as BRaTS 2021 [3]. It utilizes shifted windows for self attention at five different resolutions, allowing the model to learn more complex relations within the scan.

3   Results

Table 1, 2, and 3 display multiple metrics used to evaluate our architectures. For each metric value we included the 95% confidence interval (CI) in parentheses.

Table 1. Internal testing set primary tumor (label 1) results with 95% CI Note: CI is confidence interval, and values in parentheses are the CIs.

Model	Dice score	Jaccard	Median recall	Median precision
3D nnU-Net	**0.821 (0.6937, 0.7923)**	**0.697 (0.5737, 0.6759)**	0.823 (0.6949, 0.8082)	0.886 (0.7794, 0.8661)
nnM-Net	0.814 (0.5522, 0.6987)	0.687 (0.5549, 0.6628)	0.765 (0.6498, 0.7673)	**0.895 (0.7710, 0.8754)**
SwinUNet	0.8047 (0.7426, 0.7581)	0.6732 (0.4688, 0.5972)	**0.8328 (0.6674, 0.8022)**	0.7879 (0.5841, 0.7313)

Table 2. Internal testing set lymph node (label 2) results with 95% CI

Model	Dice score	Jaccard	Median recall	Median precision
3D nnU-Net	**0.74 (0.5944, 0.6981)**	**0.587 (0.3305, 0.4606)**	0.647 (0.4557, 0.6265)	**0.684 (0.6120, 0.7365)**
nnM-Net	0.698 (0.5693, 0.6742)	0.536 (0.3237, 0.4500)	**0.671 (0.4587, 0.6327)**	0.654 (0.6048, 0.7228)
SwinUNet	0.6690 (0.4524, 0.6053)	0.5026 (0.2613, 0.3905)	0.6504 (0.4877, 0.6494)	0.6761 (0.4190, 0.5969)

Table 3. Internal testing set averaged results

Model	Dice score	Jaccard	Median recall	Median precision
3D nnU-Net	**0.7805 (0.5797, 0.6776)**	**0.642 (0.4676, 0.5578)**	0.735 (0.5963, 0.7026)	**0.785 (0.7193, 0.7961)**
nnM-Net	0.756 (0.5707, 0.6687)	0.6115 (0.4570, 0.5470)	0.718 (0.5780, 0.6837)	0.7745 (0.7123, 0.7945)
SwinUNet	0.7369	0.5879	**0.7416**	0.732

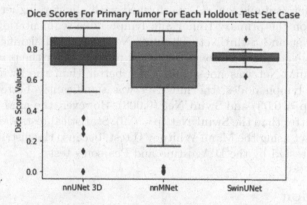

Fig. 4. Box and whisker plot of primary tumor Dice scores for each test case on our three submission models.

To calculate aggregated Dice scores (DSC), we utilized the following formula as described in the HECKTOR Challenge Website (Fig. 4).

$$DSC_{agg} = \frac{2\sum_i^N \sum_k \hat{y}_{i,k} y_{i,k}}{\sum_i^N \sum_k (\hat{y}_{i,k} + y_{i,k})} \tag{1}$$

For both the segmentation of GTVp and GTVn, we accumulate intersections and unions between the ground truth and predicted volumes across images, then divide the aggregated intersection by the aggregated union for both primary and node. Lastly, we compute the average of the two aggregated Jaccard indices. We

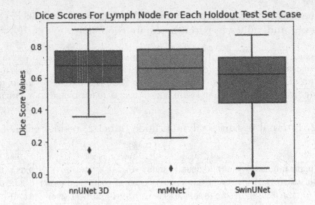

Fig. 5. Box and whisker plot of lymph node Dice scores for each test case on our three submission models.

compute the aggregated DSC separately for primary and node [5]. Our aggregated Dice score script can be found at our public Github repository at https://github.com/xmuyzz/HECKTOR2022 (Fig. 5).

The nnU-Net (submission #1) performed with slightly higher Dice score, Jaccard index on the primary tumor and lymph node segmentation tasks than did the nnM-Net and SwinUNet. The nnU-Net performed significantly better than the SwinUNet ($p = 0.03$) on the internal testing set for the primary tumor, however, the nnM-Net was not significantly better than the SwinUNet ($p = 0.14$). For the lymph nodes, the nnU-Net was significantly better than both the nnM-Net ($p = 0.03$) and SwinUNet (0.008). However, the nnM-Net was not significantly better than the SwinUNet ($p = 0.9$). Statistical tests were performed on the testing set using the Mann Whitney U test, because the distributions were not normal as tested by the D'Agostino and Pearson's test.

4   Discussion

In this study, we developed and validated three model frameworks with goal of optimizing a medical imaging auto-segmentation pipeline for head and neck cancer. We found that each model demonstrated promising performance, with average Dice scores of at least 0.73 for combined tumor and lymph node prediction.

We found that 3D-based models outperformed 2D models, and that use of the nnU-Net framework on combined CT and PET data seemed to perform well. Future research should examine better use of the PET data such as attention mechanisms, more sophisticated cropping, normalization, or thresholding. Ensembling of multiple state-of-the-art models may also yield improved results and must be further tested.

5 Conclusion

In this work, we investigate the utilization of the nnU-Net, MNet, and Swin Transformer for automatic segmentation of primary tumors and lymph nodes. Our robust evaluation results demonstrate the effectiveness of utilizing automated methods for tumor and lymph node segmentation. It highlights the robustness of state-of-the-art machine learning segmentation for this task through the evaluation on multi-institutional data. Future work would aim at fine-tuning network architecture parameters to even better exploit their potential to segment images.

References

1. Andrearczyk, V., et al.: Overview of the HECKTOR challenge at MICCAI 2021: automatic head and neck tumor segmentation and outcome prediction in PET/CT images. In: Andrearczyk, V., Oreiller, V., Hatt, M., Depeursinge, A. (eds.) HECKTOR 2021. LNCS, vol. 13209, pp. 1–37. Springer, Cham (2022). https://doi.org/10.1007/978-3-030-98253-9_1
2. Dong, Z., et al.: MNet: rethinking 2D/3D networks for anisotropic medical image segmentation. arXiv preprint arXiv:2205.04846 (2022)
3. Hatamizadeh, A., Nath, V., Tang, Y., Yang, D., Roth, H., Xu, D.: Swin UNETR: swin transformers for semantic segmentation of brain tumors in MRI images (2022). https://doi.org/10.48550/ARXIV.2201.01266
4. Isensee, F., Jaeger, P.F., Kohl, S.A., Petersen, J., Maier-Hein, K.H.: nnU-Net: a self-configuring method for deep learning-based biomedical image segmentation. Nat. Methods 18(2), 203–211 (2021)
5. Kumar, N., Verma, R., Sharma, S., Bhargava, S., Vahadane, A., Sethi, A.: A dataset and a technique for generalized nuclear segmentation for computational pathology. IEEE Trans. Med. Imaging 36(7), 1550–1560 (2017). https://doi.org/10.1109/TMI.2017.2677499
6. Oreiller, V., et al.: Head and neck tumor segmentation in PET/CT: the HECKTOR challenge. Med. Image Anal. 77, 102336 (2022). https://doi.org/10.1016/j.media.2021.102336
7. Vigneswaran, N., Williams, M.D.: Epidemiologic trends in head and neck cancer and aids in diagnosis. Oral Maxillofac. Surg. Clin. North Am. 26(2), 123–141 (2014)
8. Ye, Z., et al.: Deep learning-based detection of intravenous contrast enhancement on CT scans. Radiol. Artif. Intell. 4(3), e210285 (2022)
9. Yeung, M., Sala, E., Schönlieb, C.B., Rundo, L.: Unified focal loss: generalising dice and cross entropy-based losses to handle class imbalanced medical image segmentation (2021). https://doi.org/10.48550/ARXIV.2102.04525

Fusion-Based Automated Segmentation in Head and Neck Cancer via Advance Deep Learning Techniques

Seyed Masoud Rezaeijo[1], Ali Harimi[2], and Mohammad R. Salmanpour[3,4,5]($\boxtimes$)

[1] Department of Medical Physics, Faculty of Medicine, Ahvaz Jundishapur University of Medical Sciences, Ahvaz, Iran

[2] Department of Electrical Engineering, Shahrood Branch, Islamic Azad University, Shahrood, Iran

[3] Technological Virtual Collaboration (TECVICO Corp), Vancouver, BC, Canada
msalman@bccrc.ca

[4] Department of Physics and Astronomy, University of British Columbia, Vancouver, BC, Canada

[5] Department of Integrative Oncology, BC Cancer Research Institute, Vancouver, BC, Canada

Abstract. Background: Accurate prognostic stratification and segmentation of Head-and-Neck Squamous-Cell-Carcinoma (HNSCC) patients can be an important clinical reference when designing therapeutic strategies. We set to automatically segment HNSCC using advanced deep learning techniques linked to the image fusion technique.

Method: 883 subjects were extracted from HECKTOR-Challenge. 524 subjects were considered for the training and validation procedure, and 359 subjects as external testing were employed to validate our segmentation models. First, PET images were registered to CT images. The resultant images are cropped after an enhancement procedure. Subsequently, a weighted fusion technique was employed to combine PET and CT information. To this end, we developed a Cascade-Net consisting of two states of art neural networks to segment the tumors via the fused image. Our segmentation framework performs in three main stages. In the first stage, which is an organ localizer module, a candidate segmentation region of interest (ROIs) for each organ is generated. The second stage is a 3D U-Net refinement organ segmentation which produces a more robust and accurate contour from the previous coarse segmentation mask. This network is equipped with an attention mechanism on skip connections and a deep supervision concept that generates ROIs by eliminating irrelevant background information. This network will identify the probability of the presence of each organ. In the last stage, the extracted regions will be fed to the 3D ResU-Net to generate a fine segmentation. The performance of the proposed framework was evaluated through well-established quantitative metrics such as the dice similarity coefficient.

Result: Using the weighted fusion technique linked with Cascade-Net, our method provided the average dice score of 0.71. Moreover, this algorithm resulted in dice score of 0.74, and 0.68 for the primary gross tumor volume (GTVp) and metastatic nodes (GTVn), respectively.

Conclusion: We demonstrated that using the fusion technique followed by an appropriate automatic segmentation technique provides a good performance.

V. Andrearczyk et al. (Eds.): HECKTOR 2022, LNCS 13626, pp. 70–76, 2023.
https://doi.org/10.1007/978-3-031-27420-6_7

Keywords: Head and neck squamous cell carcinoma · Deep learning · Segmentation techniques · Fusion techniques

1 Introduction

Head and Neck (H&N) cancers, involving more than a half million people each year with a death rate of 50% [1], have been considered the sixthmost highly incident cancer worldwide. Radiotherapy and target delineation is the most popular treatment option. Automatic segmentation of H&N tumors in the initial stages is crucial since manual inspections are time-consuming, tough, and highly related to the experience of radiologists, which increases the risk of misunderstanding. Early diagnosis of brain tumors plays an essential role in improving treatment possibilities [2]. Medical imaging modalities such as Positron Emission Tomography (PET), Single-Photon Emission Computed Tomography (SPECT), Computed Tomography (CT), Magnetic Resonance Spectroscopy (MRS), and Magnetic Resonance Imaging (MRI) can be used to provide helpful information about shape, size, location, and metabolism of brain tumors [3]. The position and boundary of normal tissues can also be observed in CT images [4].

Although PET is fruitful in early disease detection [5], automatic H&N tumor segmentation based on PET images is challenging because these images lack spatial resolution [6] and have a high intrinsic noise level. Furthermore, deep learning-based gross tumor volume (GTV) auto segmentation algorithms indicated that PET plays a more crucial role than CT in GTV segmentation [7]. This is attributable to the fact that PET images highlight high-metabolic activities from tumor cells. However, the limited spatial resolution and low signal-to-noise ratio of FDG-PET images [6] can lead to erroneous predictions. It is necessary to mention that PET images also have higher intensity variations between different institutions than CT images, which can be caused by differences in individual FDG doses and scanner settings. Hence, the fusion of both PET and CT image modes provides a more informative representation for automated segmentation of tumors and target delineation [5, 6, 8–10].

Since different image modalities such as MRI, ultrasound, and PET include specific information (perspectives) [9, 11–13] of the same object, image fusion techniques enable us to combine two or more images to enhance the information content, improve the performance of object recognition systems by integrating many sources of imaging systems, help in sharpening the images, improve geometric corrections, enhance certain features that are not visible in either of the images, replace the defective data, complement the data sets for better decision making, and keep the integrity of important features or remove the noise [11]. It has been shown that fusing different images reduces ambiguity and enhances the reliability of defect detection in both visual and qualitative evaluation [14]. As such, many studies [5, 8–10, 12, 15–17] indicated that fusion techniques are considered a vital pre-processing phase for several applications such as outcome prediction, early diagnosis, segmentation (or delineation), and others.

Annual HECKTOR challenges [13, 18], starting in 2020 first, encouraged participants to develop automatic segmentation of H&N tumors using PET and CT modalities. In the past few years, deep learning (DL) techniques based on convolutional neural networks (CNN) delivered excellent medical image analysis tasks, including image segmentation. Wang et al. [19] used 3D U-Net as the backbone architecture, on which a residual network is added to better capture image detail information. Xie et al. [20] adopted the 3D U-Net network to carry out automated segmentation of H&N tumors based on the dual modality PET-CT images. In a study [5], 3DUNETR (UNet with Transformers) and 3D-UNet have been utilized to segment HNSCC tumors automatically. In our previous studies [9], multiple fusion techniques were utilized to combine PET and CT information. Moreover, we employed 3D-UNet architecture and SegResNet (segmentation using autoencoder regularization) to improve segmentation performance.

In this study, as elaborated next, we first registered PET to CT, then cropped those. Subsequently, we employed weighted fusion to combine PET and CT information. Moreover, we proposed a coarse segmentation map model resulting from a cascade attention U-Net model trained using the fused image. The map is then linked with input data to fine-tune the model further. Section 2 details the proposed method. Experimental results, analysis, and discussion are provided in Sect. 3. Finally, the paper concludes in Sect. 4.

2 Materials and Methods

2.1 Dataset, PET/CT Acquisition

We collected 683 datapoints with PET and CT images from the HECKTOR challenge dataset (MICCAI 2022). 524 subjects were considered for the training and model selection, and 359 subjects were employed for external testing of selected model. 3D ground truth of training datapoints was given by the HECKTOR challenge. In the pre-processing step, all PET images were first registered to CT images using rigid registration algorithm. Image registration is the method of aligning images so relationships between them can be seen more quickly. The same term is also applied to explain the alignment of images to a computer model or physical space. Image registration was performed using six degrees of freedom, consisting of three translations and three rotations based on patient-wise registration. The fusion techniques combine and convert sole CT and sole PET images to a single image. Hence, the fusion technique enables the exploitation of the powers of all imaging modalities together, reducing or minimizing the faults of every single modality. Recent studies [5, 9, 10, 12, 13, 15–17] indicated that weighed transform algorithm outperformed other fusion techniques some studies. Therefore, weighed transform algorithm, developed as in-house-developed Python code, was employed to fuse PET and CT images. Figure 1 demonstrates some examples of CT, PET, and fused images.

2.2 Proposed Deep Learning Algorithm

In this study, we proposed a twice attention-based framework comprised of two cascade U-Net models with a channel attention block in the Encoder and a spatial attention block in the decoder [20]. In the first stage, the network generates a coarse segmentation map, which is applied to the second network after concatenating images. The employed image of this stage is applied to the network in two channels; the first channel consists of a

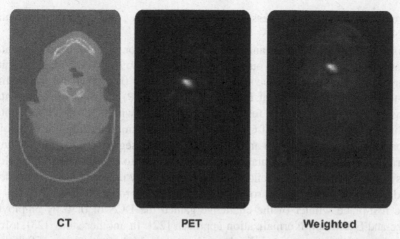

CT   PET   Weighted

Fig. 1. Examples of 1) CT image alone, 2) PET image alone, 3) the fused image generated by weighted fusion technique

fusion of PET and CT images with weights of 0.9 and 0.1, respectively. The images of the second channel are generated from the fusion of PET and CT images with weights of 0.8 and 0.2, respectively.

2.3 Analysis Procedure

First, all PET images are resampled to the size of CT images, developed as in-house-developed Python code. After cropping images, the weighted fusion technique is employed to combine these two images. In cropping prosecute, morphological operators are employed to remove all redundancy, including areas outside the featured space. Such space is non-informative in terms of existing textures and anatomical information. Then, the data augmentation is performed to increase the data size and robustness of the algorithm. It includes flip left/right, random rotation, and random cropping. It also prevents the network from overfitting by increasing the model generalization. The dice loss is maximized using the Adam optimizer in the binary cross entropy function. The network is trained by a learning rate of 0.0001 in 600 epochs. We used sole PET image in the first step to train the network. Next, the PET images were fused with CT images in a weighted manner and the fuse images were applied to the network. Finally, the model is trained using a cascade structure, more elaborated above. We reported dice score (DCS) to compare the models.

3   Results and Discussion

We applied two deep learning algorithms for segmentation on CT, PET, and the fused image generated form fusion techniques. Using the weighted fusion technique linked with Cascade-Net, our method provided the average DCS of 0.71. Moreover, this algorithm resulted in DCSs of 0.74, and 0.68 for the GTVp and GTVn, respectively. Our results confirmed the effectiveness of the employed U-Net model for the segmentation of H&N tumors.

This study demonstrated enhanced task achievement for the automated segmentation of tumors in task 1, compared to our previous studies [5, 9, 20]. At the same time, in our past studies [5], we achieved dice scores around 0.63 and 0.68 through LP-SR Mixture fusion technique (the mixture of Laplacian Pyramid (LP) and Sparse Representation (SR) fusion techniques) and LP fusion technique linked with a 3D-UNet segmentation algorithm. In short, the usage of Cascade-Net segmentation algorithm, instead of 3D-UNet based segmentation led to an improvement in the segmentation of H&N tumors. Furthermore, automated segmentation has the potential to allow radiation oncologists or physicians to improve treatment planning efficiency by reducing the time needed for tumor delineation as well as improving inter-observer reproducibility [21].

The first-place winner of the challenge gained the DCS of 0.76 by employing a Squeeze-and-Excitation Normalization approach [22]. In another study [23], a framework with iterative refinement ability has achieved the highest precision of 0.85 for the segmentation of H&N in this challenge. In HECKTOR 2021, the winner of the challenge achieved the dice score of 0.78 using a U-Net model [20]. Overall, we believe the usage of the fusion technique and pre-processing method may enable the researchers to improve the segmentation performance [5, 8–10, 16].

As a limitation in this study, future studies with a large sample size are suggested. Our study considered automated segmentation and survival prediction for two-images CT and PET; hence the proposed approaches can also be used for other related tasks in medical image analysis such as mpMR images, including T2 weighted image (T2W), diffusion-weighted magnetic resonance imaging (DWI), apparent diffusion coefficient (ADC), and dynamic contrast-enhanced magnetic resonance imaging (DCE-MRI). Moreover, this study can further benefit from more fusion techniques to combine PET and CT images. In addition, future studies should seek to employ more deep learning segmentation algorithms and further optimize these methods for improved H&N tumor segmentation performance in forthcoming iterations of the HECKTOR Challenge. The novelty of this study is the usage of a fusion technique to generate new images from CT and PET images.

4   Conclusions

This effort indicated the development and validation of deep learning model, a twice attention-based framework comprised of two cascade U-Net models with a channel attention block in the Encoder and a spatial attention block in the decoder, to segment H&N tumors in an end-to-end automated workflow based on PET/CT images and the fused image. Using a combination of pre-processing steps, and the deep learning algorithm, we achieve an average dice score of ~71% for our best model. Our method notably improves upon the previous iteration of our model submitted in the 2021 HECKTOR Challenge. Overall, we showed that the usage of image fusion followed by appropriate automated segmentation provides good performance, compared to our previous studies.

Acknowledgement. This study was supported by Technological Virtual Collaboration Corporation (TECVICO Corp.) located in Vancouver, BC, Canada.

Code Availability. All codes are publicly shared at: https://github.com/Tecvico.

Conflict of Interest. The authors have no relevant conflicts of interest to disclose.

References

1. Wu, Z.-H., Zhong, Y., Zhou, T., Xiao, H.-J.: miRNA biomarkers for predicting overall survival outcomes for head and neck squamous cell carcinoma. Genomics **113**(1), 135–141 (2021)
2. Butowski, N.A.: Epidemiology and diagnosis of brain tumors. CONTINUUM: Lifelong Learn. Neurol. **21**(2), 301–313 (2015)
3. Drevelegas, A., Papanikolaou, N.: Imaging modalities in brain tumors. In: Drevelegas, A. (ed.) Imaging of Brain Tumors with Histological Correlations, pp. 13–33. Springer, Heidelberg (2011). https://doi.org/10.1007/978-3-540-87650-2_2
4. Wang, J., Peng, Y., Guo, Y., Li, D., Sun, J. (2022). CCUT-Net: pixel-wise global context channel attention UT-Net for head and neck tumor segmentation. In: Andrearczyk, V., Oreiller, V., Hatt, M., Depeursinge, A. (eds.) HECKTOR 2021. LNCS, vol. 13209, pp. 38–49. Springer, Cham. https://doi.org/10.1007/978-3-030-98253-9_2
5. Fatan, M., Hosseinzadeh, M., Askari, D., Sheikhi, H., Rezaeijo, S.M., Salmanpour, M.R.: Fusion-based head and neck tumor segmentation and survival prediction using robust deep learning techniques and advanced hybrid machine learning systems. In: Andrearczyk, V., Oreiller, V., Hatt, M., Depeursinge, A. (eds.) HECKTOR 2021. LNCS, vol. 13209, pp. 211–23. Springer, Cham (2022). https://doi.org/10.1007/978-3-030-98253-9_20
6. Rahmim, A., Zaidi, H.: PET versus SPECT: strengths, limitations and challenges. Nucl. Med. Commun. **29**(3), 193–207 (2008)
7. Ren, J., Eriksen, J.G., Nijkamp, J., Korreman, S.S.: Comparing different CT, PET and MRI multi-modality image combinations for deep learning-based head and neck tumor segmentation. Acta Oncol. **60**(11), 1399–1406 (2021)
8. Javanmardi A, Hosseinzadeh M, Hajianfar G, Nabizadeh AH, Rezaeijo SM, Rahmim A, et al. Multi-modality fusion coupled with deep learning for improved outcome prediction in head and neck cancer. In: Medical Imaging 2022: Image Processing, pp. 664–668. SPIE (2022)
9. Salmanpour, M.R., Hajianfar, G., Rezaeijo, S.M., Ghaemi, M., Rahmim, A.: Advanced automatic segmentation of tumors and survival prediction in head and neck cancer. In: Andrearczyk, V., Oreiller, V., Hatt, M., Depeursinge, A. (eds.) HECKTOR 2021. LNCS, vol. 13209, pp. 202–10. Springer, Cham (2022). https://doi.org/10.1007/978-3-030-98253-9_19
10. Salmanpour, M.R., Hosseinzadeh, M., Modiri, E., Akbari, A., Hajianfar, G., Askari, D., et al.: Advanced survival prediction in head and neck cancer using hybrid machine learning systems and radiomics features. In: Medical Imaging 2022: Biomedical Applications in Molecular, Structural, and Functional Imaging, pp. 314–321. SPIE (2022)
11. Taxak, N., Singhal, S.: A Review of image fusion methods. Int. J. Innovative Sci. Res. Tech. **8**(3), 598–601 (2018)
12. Rezaeijo, S.M., Hashemi, B., Mofid, B., Bakhshandeh, M., Mahdavi, A., Hashemi, M.S.: The feasibility of a dose painting procedure to treat prostate cancer based on mpMR images and hierarchical clustering. Radiat. Oncol. **16**(1), 1–16 (2021)
13. Andrearczyk, V., et al.: Overview of the HECKTOR challenge at MICCAI 2021: automatic head and neck tumor segmentation and outcome prediction in PET/CT images. In: Andrearczyk, V., Oreiller, V., Hatt, M., Depeursinge, A. (eds.) HECKTOR 2021. LNCS, vol. 13209, pp. 1–37. Springer, Cham (2022). https://doi.org/10.1007/978-3-030-98253-9_1
14. Wang, Q., Shen, Y., Jin, J.: Performance evaluation of image fusion techniques. Image Fusion: Algorithms Appl. **19**, 469–492 (2008)

15. Salmanpour, M.R., Hajianfar, G., Lv, W., Lu, L., Rahmim, A.: Multitask Outcome Prediction using Hybrid Machine Learning and PET-CT Fusion Radiomics. Soc. Nucl. Med. **62**(supplement 1), 1424 (2021)

16. Salmanpour, M.R., Hosseinzadeh, M., Akbari, A., Borazjani, K., Mojallal, K., Askari, D., et al.: Prediction of TNM stage in head and neck cancer using hybrid machine learning systems and radiomics features. In: Medical Imaging 2022: Computer-Aided Diagnosis, pp. 648–53. SPIE (2022)

17. Salmanpour, M.R., Hosseinzadeh, M., Rezaeijo, S.M., Uribe, C., Rahmim, A.: Robustness and reproducibility of radiomics features from fusions of PET-CT images. Soc. Nucl. Med. **63**(supplement 2), 3179 (2022)

18. Andrearczyk, V., et al.: Overview of the HECKTOR challenge at MICCAI 2020: automatic head and neck tumor segmentation in PET/CT. In: Andrearczyk, V., Oreiller, V., Depeursinge, A. (eds.) HECKTOR 2020. LNCS, vol. 12603, pp. 1–21. Springer, Cham (2021). https://doi.org/10.1007/978-3-030-67194-5_1

19. Wang, G., Huang, Z., Shen, H., Hu, Z.: The head and neck tumor segmentation in PET/CT based on multi-channel attention network. In: Andrearczyk, V., Oreiller, V., Hatt, M., Depeursinge, A. (eds.) HECKTOR 2021. LNCS, vol. 13209, pp. 68–74. Springer, Cham (2022). https://doi.org/10.1007/978-3-030-98253-9_5

20. Xie, J., Peng, Y.: The head and neck tumor segmentation based on 3D U-Net. In: Andrearczyk, V., Oreiller, V., Hatt, M., Depeursinge, A. (eds.) HECKTOR 2021. LNCS, vol. 13209, pp. 92–98. Springer, Cham (2022). https://doi.org/10.1007/978-3-030-98253-9_8

21. Oreiller, V., Andrearczyk, V., Jreige, M., Boughdad, S., Elhalawani, H., Castelli, J., et al.: Head and neck tumor segmentation in PET/CT: the HECKTOR challenge. Med. Image Anal. **77**, 102336 (2022)

22. Iantsen, A., Visvikis, D., Hatt, M.: Squeeze-and-excitation normalization for automated delineation of head and neck primary tumors in combined PET and CT images. In: Andrearczyk, V., Oreiller, V., Depeursinge, A. (eds.) HECKTOR 2020. LNCS, vol. 12603, pp. 37–43. Springer, Cham (2021). https://doi.org/10.1007/978-3-030-67194-5_4

23. Chen, H., Chen, H., Wang, L.: Iteratively refine the segmentation of head and neck tumor in FDG-PET and CT images. In: Andrearczyk, V., Oreiller, V., Depeursinge, A. (eds.) HECKTOR 2020. LNCS, vol. 12603, pp. 53–58 . Springer, Cham (2021). https://doi.org/10.1007/978-3-030-67194-5_6

Stacking Feature Maps of Multi-scaled Medical Images in U-Net for 3D Head and Neck Tumor Segmentation

Yaying Shi[1]([✉]), Xiaodong Zhang[2], and Yonghong Yan[1]

[1] University of North Carolina at Charlotte, Charlotte, NC, USA
yshi10@uncc.edu
[2] The University of Texas MD Anderson Cancer Center, Houston, TX, USA

Abstract. Machine learning, especially deep learning, has achieved state-of-the-art performance on various computer vision tasks. For computer vision tasks in the medical domain, it remains as challenging tasks since medical data is heterogeneous, multi-level, and multi-scale. Head and Neck Tumor Segmentation Challenge (HECKTOR) provides a platform to apply machine learning techniques to the medical image domain. HECKTOR 2022 provides positron emission tomography/computed tomography (PET/CT) images which includes useful metabolic and anatomical information to sufficiently make an accurate tumor segmentation. In this paper, we proposed a stacked-multi-scaled medical image segmentation framework to automatically segment the Head and Neck tumor using PET/CT images. The main idea of our network was to generate various low-resolution feature maps of PET/CT images to make a better contour of Head and Neck tumors. We used multi-scaled PET/CT images as inputs, and stacked different intermediate feature maps by resolution for a better inference result. In addition, we evaluated our model on the HECKTOR challenge test dataset. Overall, we achieved a 0.69786, 0.66730 mean Dice score on GTVp and GTVn respectively. Our team's name is HPCAS.

Keywords: Medical image segmentation · HECKTOR challenge · Machine learning · Convolutional neural network · Deep learning

1 Introduction

Head and Neck (H&N) cancer is one of the most common cancers which includes several positions on your throat, nose and other head areas except brain and eyes. There is an estimated of 277,597 people died from H&N cancer in the worldwide [4]. The 5-year survive rate of H&N cancer is around 90% if the cancer was detected in stage 1 [3], which significantly reduces to 70% in stage 2. The survival rate is 60% and 30% for stage 3 and stage 4 respectively [3]. Early diagnoses and treatments of H&N cancer can improve the survival rate of patients. Medical images such as positron emission tomography (PET)/computed tomography (CT) had shown great value to localize the primary tumor and assist physicians with contouring the tumor. However, it has been low-efficient for physician to manually contour H&N tumor slide by slide, and the distinctive size, type, and shape of the tumor has made it as a challenging task to defined an uniformed pattern [15].

An auto-segmentation method can be a feasible solution for the two problems listed above. Traditional segmentation methods such as threshold method [18], region-based method [10] and edge-base method [5] have their own limitations; and is hard to find the best threshold for the tumor. In recent, Deep Learning based auto segmentation method has gained more attention for its great potential in computer vision tasks. In the medical domain, many studies have been conducted by using Deep Learning (DL) method such as Wang [16] proposed a triple cascaded framework which can hierarchically segment three different types of brain tumors. EMMA [8] introduced a novel ensemble of DeepMedic [9], FCN [11] and Unet [14] models and get a better segmentation result for brain tumors.

However, there are limited studies work on H&N tumor segmentation until the raise of Head and Neck tumor segmentation challenge [13]. Head and Neck tumor segmentation challenge 2022 (HECKTOR) is a competition provides well-labeled H&N tumor dataset for competitors to find out the best segmentation method for H&N tumor [1]. It provides a platform for H&N tumor segmentation and good quality tumor dataset. Meanwhile, it also provides various state-of-art H&N tumor segmentation algorithms and demonstrates the potential of deep learning on H&N tumor segmentation.

Recent post-challenge proceedings in HECKTOR challenge shows that most methods are implemented on U-Net [14]. Our network was inspired from Unet with some modifications. In our work, we proposed a 3D 'U' shape network architecture which takes multi-scaled PET/CT images as an input for H&N tumor segmentation and concatenate their feature maps for deconvolution.

The remainder of the paper is organized as follows: Sect. 2 introduces related work of our method; Sect. 3 presents our proposed network structure, pre-processing methods, augmentation details, training details and so on; Sect. 4 evaluates our model with HECKTOR 2022 Challenge dataset; Sect. 5 analyzes the result and discusses future improvements.

2 Related Work

Convolution neural networks have been used as fundamental model architecture for medical image segmentation in recent years. For medical image segmentation, U-net, which is a famous network for cell segmentation, has been used as the base-line model in medical image domain [14]. U-net introduced a 'U' shape network structure, which is also known as encoder-decoder network structure. After the U-net, there are many subsequent variants of it, such as V-net [12] which extends the U-net by adding a new Dice loss function. Dice loss is widely used in many medical image segmentation methods. Some variants of U-net have modified and improved the network by adding new blocks. Residual UNet [19] improved the U-net by adding a new residual block. Densely connected Unet [6] added a new densely connect layer to the U-net. TranBTS [17] introduced a new approach by combining Unet with self-attention mechanisms, and added self-attention block at the bottom layer of 'U' shape network.

3   Methods

In this section, we will first introduce the pre-processing of raw HECKTOR challenge data, and augmentation. As we mentioned in Sect. 1, we used stacked multiple resolution input images for training. We will also provide the detailed splitting methods, training details and the loss function that we used.

3.1   Preprocessing

Based on the past challenge experiences, the dataset has no distribution shifts among different medical institutions. Since we did not perform any n-fold cross validation metrics in this challenge, we do not have to split the training dataset. Thus, we trained with whole dataset without any train-validation splits.

The PET/CT images from HECKTOR challenge vary from sizes and not registered. Thus, we first applied a resampling script to resample all the input PET/CT images, well-labeled ground truth into a registered form. The script we used was adapted from official HECKTOR challenge GitHub website. The resample space is $2.0 \times 2.0 \times 2.0$. The directions and sizes of resampled images were decided by the pair of original PET/CT image, which means after registration each training cases has different size in dimension. The interpolation for each PET/CT image is sitkBSpline from SimpleITK library [2].

Compared with the past challenge, HECKTOR 2022 did not provide any bounding boxes. Thus, we applied a random crop on resampled dataset and ensured all images are in the dimension of (144, 144, 144). If the dimension of resampled image is smaller than (144, 144, 144), we would apply a zero-padding to make sure its dimension is (144, 144, 144). As we had mentioned in Sect. 1, we used a multi-scale image as input which is multi-resolution input image. We resampled those random cropped images into the dimension of (72, 72, 72), (36, 36, 36) and (18, 18, 18).

3.2   Stacked Multi-scale 'U' Shape Network

The overall network architecture is shown as Fig. 1. It was implemented based on the U-Net. It has encoder, decoder part, bottom layer and feature maps from multi-scale input images. Since most of the medical images are 3-dimension (3D), we first extended the original U-net [14] into 3D. We did not add any blocks such as residual block, dense block, self-attention and so on to maintain simplicity of our network. The encoder part consisted of three different resolution inputs. Those inputs were registered in the same origin, spacing and direction. For low resolution PET/CT input images, two successive convolution layers without pooling were used to extract the feature maps from different scale input images. At the top, there are full size input images fed into three down-sampling blocks. Each down-sampling block consists of two convolution layers with batch normalization, Relu activation function, zero padding, and pooling layer. The decoder part contains some regular deconvolution layers. At the bottom we copied the low-resolution feature map from encoder to decoder. At decoder part, we contacted deconvolution layer output, encoder part feature maps and low resolution input feature maps. Thus, our decoder part contains more rich feature information.

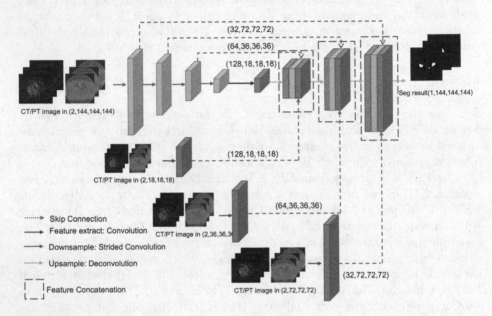

Fig. 1. Stacked multi-scaled input 'U' Network Architecture. We used an input patch size of 144 × 144 × 144, 72 × 72 × 72, 36 × 36 × 36, 18 × 18 × 18 with PET/CT as two modalities for the network. The network structure basically is U-shape architecture which was implemented based on U-net. The down-sampling is implemented with three down blocks which has a strided 3D convolution operation with a 3 × 3 × 3 filter for each modality. The up-sampling is done with deconvolution. The size of feature map was displayed in the figure. We directly copied the feature maps at bottom layer. We concatenated different resolution input image feature maps with deconvolution output. We also used skip connection to directly concatenate feature maps from encode part.

From the left side of Fig. 1, the input has 2 channeled 3D images which are CT and PET image. The first level of input image size is 144 × 144 × 144. In the encoder part, it fed into down-sampling block for feature extraction. The initial filter size for convolution layer was set as 32. At the bottleneck, we mirrored the feature maps from encoder to decoder part. For low-resolution input at second level such as (72, 72, 72), it went through two convolution layers with an initial filter size as 32. When we have all the three low resolution images' feature maps, we will concatenate it with same resolution output in the decode part. The learning rate of our model was set as 2e−4 and it was reduced as the formula:

$$Lr = Lr * \left(1 - \frac{e}{te}\right)^{0.9} \qquad (1)$$

where e is current epoch, te is total number of epochs. We used L2 norm regularization on the convolutional kernel parameters with a weight of 1e−5. We ran the training for 2000 epochs in total.

3.3 Optimization and Data Augmentation

nnUnet [7] provides some helpful guidelines for medical image segmentation. In this paper, we applied some of data augmentations and optimizations according to the suggestion of nnUnet.

Since we did not use n-fold cross-validation technique, we do not have any validation set. Thus, we do not have an early stopping mechanism nor ensembling strategies.

We increased the batch size from 2 (original Unet paper) to a size of 16. According to the description of nnUnet, lower batch size will generate unnecessary noisier gradients. Those noisier gradient can reduce the over-fitting issue but decreases the overall performance. So, we used a higher batch size for better performance and less data copy between the host and the device.

We also applied some data augmentation techniques to get a more robust model from the training. In the pre-processing section, we had already applied a random crop and zero-padding for the input dataset. We also used a Z-score normalization on the input PET/CT images. Besides, we also applied a random mirror flipping method. The flipping was operated to three dimension across the axial, sagittal and coronal planes. The random rate of flipping was set as 0.5. We applied a random rotation to training dataset at ratio of 0.5. The rotation angle ranges from −10 to 10. We applied a random intensity shift at ratio of 0.5. Then intensity of training data shifted between −0.1 and 0.1. The scale of training data was set from 0.9 to 1.1.

We choose the Dice loss function as the evaluation metric during training. For the optimizer, we choose Adam optimizer. We also used batch normalization instead of instance normalization. Batch normalization can reduce the performance difference between training dataset and testing dataset. By applying previous data augmentation techniques, our best effort was implemented to reduce the domain gap among the training and testing data.

4 Result

4.1 HECKTOR 2022 Datasets

The HECKTOR 2022 datasets contains 359 testing cases and 524 training cases. In the HECKTOR training dataset, each training case consists of a set of PET/CT images and a well-labeled ground truth image. All three images are in NIfTI format. There are two types of tumors: H&N primary tumors(GTVp) and H&N nodal Gross Tumor Volumes(GTVn), which are labeled as 1 and 2 respectively. Our model was trained with the HECKTOR training dataset. We did not use any other public or private datasets. In this work, there are four different resolution of PET/CT images. The dimension of those images are shown in the last section. All of those different resolution images have two input channels. The ground truth labels have 3 classes: label 0 for the background, label 1 for GTVp tumor, label 2 for GTVn. The test dataset consists of 319 cases with no ground truth provided. Since we did not use cross-validation, we can only evaluate the results on test dataset. For the evaluation metrics, we used Dice score which was provided by HECKTOR 2022 challenge.

4.2   Implementation Details

Our model was run by PyTorch 1.9.0 with CUDA 11.1. The python version is 3.8.5, and was trained from scratch on a server with 4 NVIDIA A100 GPUs (40GB VRAM). The total number of training epochs was 2000. The batch size was set to 16. As mentioned in previous section, we applied pre-processing techniques and some data augmentations to the original training data. For inference on testing data, we applied the sliding windows inference and the size of window was set as $(144, 144, 144)$. In the first step, we used same pre-processing procedure as the training stage. Then we cropped the input data into small pieces by dimension of $(144, 144, 144)$, and we also performed the same resample method on those small pieces into low resolution. If the resample image dimension can not be cropped into an integer number of pieces, we will extend the last piece to the dimension of $(144, 144, 144)$. For example, if an input image is in dimension of $(160, 160, 160)$, we will crop it to small pieces of numpy array $(:144,:144,:144), (:144,:144, 16:160), (:144, 16:160,:144), (16:160,:144,:144), (:144, 16:160, 16:160), (16:160,:144, 16:160), (16:160, 16:160, 16:160)$. Then we concatenated the results together. For the challenge, we also applied test time augmentation which apply augmentations to different batch test data and merge predictions for inference stage.

4.3   Hector 2022 Test Result

In terms of challenge, we have this section specially listed for the competition results. Since we did not have validation dataset, there is no validation result. The test performance was copied from official HECKTOR challenge website. Since there are no other information provided, we can not analyze our results. The performance of two submissions in our model on Hector 2022 testing data are reported as Table 1.

Table 1. Dice score and on HECKTOR 2022 validation dataset. GTVp, GTVn present tumors H&N primary tumors and H&N nodal Gross Tumor Volumes respectively.

Train dataset	Dice		
	GTVp	GTVn	Mean
First Submission	0.69786	0.66730	0.68258
Second Submission	0.68610	0.66482	0.67546

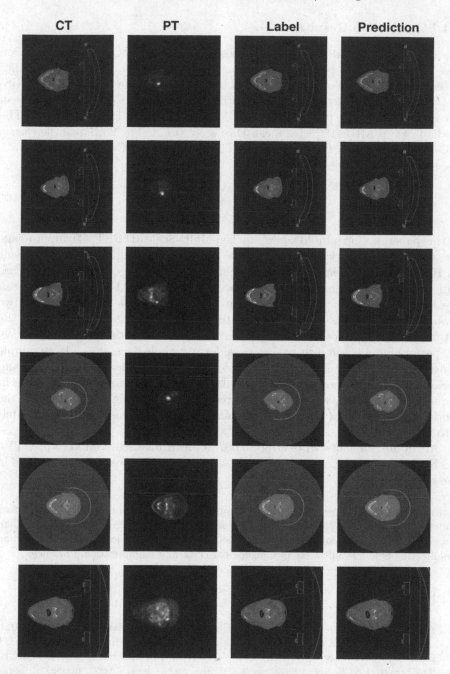

Fig. 2. Visualization of Qualitative Results. For each row, PT image is shown on the first left column. The second left is CT image. Label is next to the CT image. The predicted outcome is on the last right column. GTVp is shown in green, GTVn in red. From the first row to the last row, we displayed the best, 75th quantile, mean, median, 25th quantile, and worst validation case respectively. (Color figure online)

4.4   Qualitative Results

We randomly selected 100 training cases as validation dataset. The validation case were selected as best, worst, mean, median and 75th and 25th percentiles based on their Dice scores. The result shown as Fig. 2. The best cases is CHUS-094 with mean Dice score 0.960. The 75th quantile case is CHUS-040 with mean Dice score 0.880. The mean case is CHUM-015 with mean Dice score 0.678. The median case is MDA-185 with mean Dice score 0.778. The 25th quantile case is MDA-180 with mean Dice score 0.53. The worst case is CHUS-028 with mean Dice score 0.223.

5   Discussion

In this paper, we introduced a new Stacked Multi-Scale 3D PET/CT input image for 'U' Shape Network model to achieve a more promising segmentation result on H&N tumor. In testing phase, we achieved an overall good result. There are still plenty much spaces we can improve for the future. For instance, we can use a batch dice rather than an instant mini dice. In the current method, we evaluated the dice loss in every mini batch which is an instant mini dice. In the future, we will use dice loss for whole dataset which considered as a batch dice. By performing a batch dice, we can consider the whole dataset as a large sample which are trained in one batch. This is a trade-off problem between bias and variations. We can also improve the results by using cross-folder validation technique. By doing that, we can keep and record the best weight, and evaluate the model during training time. After training, we can also ensemble those models into a better model.

All in all, we achieved 0.69786, 0.66730 mean dice score on GTVp, GTVn H&N tumor respectively. We will address the above proposed improvements in the next challenge.

Acknowledgment. We acknowledge support from National Science Foundation under the Grant 2015254, from the University of Texas at Anderson Cancer Center, Texas Advanced Computing Center, and Oden Institute for Computational and Engineering Sciences initiative in Oncological Data and Computational Science.

References

1. Andrearczyk, V., et al.: Overview of the HECKTOR challenge at MICCAI 2021: automatic head and neck tumor segmentation and outcome prediction in PET/CT images. In: Andrearczyk, V., Oreiller, V., Hatt, M., Depeursinge, A. (eds.) HECKTOR 2021. LNCS, vol. 13209, pp. 1–37. Springer, Cham (2022). https://doi.org/10.1007/978-3-030-98253-9_1
2. Beare, R., Lowekamp, B., Yaniv, Z.: Image segmentation, registration and characterization in R with SimpleITK. J. Stat. Softw. **86** (2018)
3. Cancer Research UK: Survival for laryngeal cancer (2020). https://www.cancerresearchuk.org/about-cancer/laryngeal-cancer/survival

4. caner.net: Head and neck cancer: Statistics (2020). https://www.cancer.net/cancer-types/head-and-neck-cancer/statistics
5. Farag, A.A.: Edge-based image segmentation. Remote Sens. Rev. 6(1), 95–121 (1992)
6. Huang, G., Liu, Z., Van Der Maaten, L., Weinberger, K.Q.: Densely connected convolutional networks. In: Proceedings of the IEEE Conference on Computer Vision and Pattern Recognition, pp. 4700–4708 (2017)
7. Isensee, F., Jäger, P.F., Full, P.M., Vollmuth, P., Maier-Hein, K.H.: nnU-Net for brain tumor segmentation. In: Crimi, A., Bakas, S. (eds.) BrainLes 2020. LNCS, vol. 12659, pp. 118–132. Springer, Cham (2021). https://doi.org/10.1007/978-3-030-72087-2_11
8. Kamnitsas, K., et al.: Ensembles of multiple models and architectures for robust brain tumour segmentation. In: Crimi, A., Bakas, S., Kuijf, H., Menze, B., Reyes, M. (eds.) BrainLes 2017. LNCS, vol. 10670, pp. 450–462. Springer, Cham (2018). https://doi.org/10.1007/978-3-319-75238-9_38
9. Kamnitsas, K., et al.: DeepMedic for brain tumor segmentation. In: Crimi, A., Menze, B., Maier, O., Reyes, M., Winzeck, S., Handels, H. (eds.) BrainLes 2016. LNCS, vol. 10154, pp. 138–149. Springer, Cham (2016). https://doi.org/10.1007/978-3-319-55524-9_14
10. Leung, T., Malik, J.: Contour continuity in region based image segmentation. In: Burkhardt, H., Neumann, B. (eds.) ECCV 1998. LNCS, vol. 1406, pp. 544–559. Springer, Heidelberg (1998). https://doi.org/10.1007/BFb0055689
11. Long, J., Shelhamer, E., Darrell, T.: Fully convolutional networks for semantic segmentation. In: Proceedings of the IEEE Conference on Computer Vision and Pattern Recognition, pp. 3431–3440 (2015)
12. Milletari, F., Navab, N., Ahmadi, S.A.: V-Net: fully convolutional neural networks for volumetric medical image segmentation. In: 2016 Fourth International Conference on 3D Vision (3DV), pp. 565–571. IEEE (2016)
13. Oreiller, V., et al.: Head and neck tumor segmentation in PET/CT: the HECKTOR challenge. Med. Image Anal. 77, 102336 (2022)
14. Ronneberger, O., Fischer, P., Brox, T.: U-Net: convolutional networks for biomedical image segmentation. In: Navab, N., Hornegger, J., Wells, W.M., Frangi, A.F. (eds.) MICCAI 2015. LNCS, vol. 9351, pp. 234–241. Springer, Cham (2015). https://doi.org/10.1007/978-3-319-24574-4_28
15. Shi, Y., Micklisch, C., Mushtaq, E., Avestimehr, S., Yan, Y., Zhang, X.: An ensemble approach to automatic brain tumor segmentation. In: Crimi, A., Bakas, S. (eds.) BrainLes 2021. LNCS, vol. 12963, pp. 138–148. Springer, Cham (2022). https://doi.org/10.1007/978-3-031-09002-8_13
16. Wang, G., Li, W., Ourselin, S., Vercauteren, T.: Automatic brain tumor segmentation using cascaded anisotropic convolutional neural networks. In: Crimi, A., Bakas, S., Kuijf, H., Menze, B., Reyes, M. (eds.) BrainLes 2017. LNCS, vol. 10670, pp. 178–190. Springer, Cham (2018). https://doi.org/10.1007/978-3-319-75238-9_16
17. Wang, W., Chen, C., Ding, M., Li, J., Yu, H., Zha, S.: TransBTS: multimodal brain tumor segmentation using transformer. arXiv preprint arXiv:2103.04430 (2021)
18. Yanowitz, S.D., Bruckstein, A.M.: A new method for image segmentation. Comput. Vis. Graph. Image Process. 46(1), 82–95 (1989)
19. Zhang, Z., Liu, Q., Wang, Y.: Road extraction by deep residual U-Net. IEEE Geosci. Remote Sens. Lett. 15(5), 749–753 (2018)

A Fine-Tuned 3D U-Net for Primary Tumor and Affected Lymph Nodes Segmentation in Fused Multimodal Images of Oropharyngeal Cancer

Agustina La Greca Saint-Esteven[1,2(✉)], Laura Motisi[1], Panagiotis Balermpas[1], and Stephanie Tanadini-Lang[1]

[1] University Hospital Zurich, University of Zurich,
Rämistr. 100, 8091 Zurich, Switzerland
agustina.lagreca@usz.ch
[2] Computer Vision Laboratory, ETH Zurich,
Sternwartstr. 7, 8092 Zurich, Switzerland

Abstract. Head and Neck (HN) cancer has the sixth highest incidence rate of all malignancies worldwide. One of the two main curative treatments for this malignancy is radiotherapy, whose delivery depends on accurate contouring of the primary tumor and affected lymph nodes among other structures. In this study, we present a transfer learning-based approach for the automatic primary tumor and lymph nodes segmentation in fused positron emission tomography (PET) and computed tomography (CT) images belonging to the HECKTOR challenge dataset. Transfer learning is performed from the Genesis Chest CT model, a publicly available 3D U-net, pre-trained on chest CT scans. Three-fold cross-validation is employed during training, so that, on each fold, two different binary segmentation models are chosen, one for the primary tumor and one for the lymph nodes. During testing, majority voting is applied. Our results show promising performance on the training and validation cohorts, while moderate performance was observed in the test cohort.

Keywords: Segmentation · HECKTOR · Head-and-neck · Oropharyngeal · PET · CT

1 Introduction

The global incidence of Head and Neck (HN) cancer almost reached 1,000,000 cases in 2020 [1], from which approximately 100,000 were oropharyngeal cancer (OPC) patients. Radiation therapy with or without concomitant chemotherapy is, together with surgery, the gold-standard treatment for HN cancer, which requires accurate tumor and lymph node delineation in positron emission tomography (PET)/computed tomography (CT) images for precise dose delivery. This task is highly time-consuming and prone to inter- and intra-observer variability

V. Andrearczyk et al. (Eds.): HECKTOR 2022, LNCS 13626, pp. 86–93, 2023.
https://doi.org/10.1007/978-3-031-27420-6_9

errors [2]. As a consequence, in the last decade, a multitude of automatic contouring methods have been proposed for HN gross tumor volume (GTV) delineation and included in commercial software [3]. Moreover, the HEad and neCK
TumOR (HECKTOR) segmentation challenge [4,5] was created, which provides
a highly curated and extensive dataset of PET/CT scans of OPC patients with
labelled primary tumor (GTVp) and lymph nodes (GTVn) with the purpose of
advancing the current state-of-the-art.

In this paper, we present our proposal for the HECKTOR challenge 2022,
which is based on transfer learning from Models Genesis Chest 3D model [6],
a 3D U-net pre-trained on lung CT. PET and CT scans are combined into
a unique input, from which a sub-volume that spans from the nose until the
lungs is extracted in a semi-automatic approach. In order to alleviate the class
imbalance (background vs. foreground) present in the dataset, different losses
are explored.

2   Methods

2.1   Study Design

Training Cohort. The training cohort provided in the challenge consisted of
524 OPC patients from 7 different institutions: Centre Hospitalier de l'Université
de Montréal (CHUM, n = 56), Centre Hospitalier Universitaire de Poitiers
(CHUP, n = 72), Centre Hospitalier Universitaire de Sherbooke (CHUS, n = 72),
Centre Hospitalier Universitaire Vaudois (CHUV, n = 53), Hôpital Maisonneuve-
Rosemont (HMR, n = 18), Hôpital Général Juif (HGJ, n = 55) and MD Anderson
Cancer Center (MDA, n = 198). During training, three-fold cross-validation was
performed, so that, on each fold, the training set was further split into training
(roughly 75%) and validation (roughly 25%) based on the institution the cases
belonged to in order to have an external validation cohort. Hence, the first fold
included 416 cases in the training set (CHUM, CHUP, CHUS, HMR and MDA)
and 109 cases in the validation set (CHUV and HGJ); the second fold consisted
of 413 training cases (CHUP, CHUS, CHUV HMR and MDA) and 111 validation
cases (CHUM and HGJ); and the third fold included 415 training cases (CHUP,
CHUS, HGJ, HMR and MDA) and 108 cases in the validation cohort (CHUM
and CHUV).

Test Cohort. The test dataset consisted of 359 patients from 3 different insti-
tutions, with only one of them being also present in the training set: Centre
Henri Becquerel (CHB, n = 58), MDA (n = 200) and UniversitätsSpital Zürich
(USZ, n = 101).

2.2   Image Preprocessing

CT, PET scans and, when available, ground truth (GT) segmentation masks
were resampled to a resolution of $2 \times 2 \times 2$ mm and a sub-volume of $96 \times 96 \times 96$

pixels was cropped on each scan, which extended from the nasal columella to approximately the lungs' apices. To do that, the most inferior axial slice of the head that intersected with the most anterior slice was selected as the starting axial slice (Fig. 1). This slice and the following 95 caudal axial slices were included in the sub-volume and subsequently cropped around the center of mass of the starting axial slice to have a square 96 × 96 pixels shape. Manual corrections were applied when necessary. The CT Hounsfield units (HU) range was clipped between $[-100, 200]$ after which min-max normalization was applied. Min-max normalization per patient was applied for the PET scans. The 3D input to the network was built by multiplying both scans.

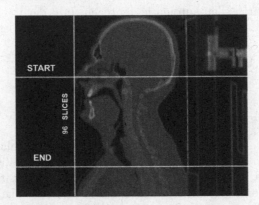

Fig. 1. Sub-volume selection in a head-and-neck CT scan.

2.3   Deep Learning Model

Genesis Chest CT model from Models Genesis [6] was fine-tuned to perform the segmentation task. The network, which is 53-layers-deep and composed of approximately 16M parameters, was originally trained by Zhou et al. as an identity auto-encoder on unlabeled chest CT scans in a self-supervised fashion.

Due to memory constraints, on each fold of the 3-fold cross-validation, two sets of binary models were created: one set for the segmentation of the GTVp and one set for the segmentation of the GTVn. On each set, three different losses (Dice coefficient loss [7], Tversky loss [8] and focal Tversky loss [9]) and three different learning rates (0.001, 0.0001 and 0.00001) were tested independently to find the optimal model. The Dice coefficient loss is a modification of the Dice similarity coefficient (DSC, Eq. 1) metric so that it becomes a minimizable objective function. $y_{i,k}$ is the ground truth (GTVp or GTVn) for voxel k and image i and $\widehat{y_{i,k}}$ is the predicted segmentation. Similarly, the Tversky loss is based on the Tversky Index (TI, Eq. 2), which is a generalization of the DSC, with the addition of two parameters, α and β, which results in a heavier penalization of false negatives. Finally, the focal Tversky loss is a generalization of the Tversky loss with the parameter γ controlling the non-linearity of the loss (Eq. 3) and forcing the model to focus on harder samples.

$$DSC = \frac{2 * \widehat{y}_{i,k} y_{i,k}}{2 * \widehat{y}_{i,k} y_{i,k} + \widehat{y}_{i,k} \overline{y}_{i,k} + \overline{\widehat{y}}_{i,k} y_{i,k}} \tag{1}$$

$$TI = \frac{\widehat{y}_{i,k} y_{i,k}}{\widehat{y}_{i,k} y_{i,k} + \alpha \overline{\widehat{y}}_{i,k} y_{i,k} + \beta \widehat{y}_{i,k} \overline{y}_{i,k}}, \alpha + \beta = 1 \tag{2}$$

$$FTL = (1 - TI)^{\gamma} \tag{3}$$

The two models with the highest validation aggregated Dice score (Eq. 4) for the GTVp and GTVn, respectively, of each fold were selected. After training, the three best models (one from each fold) were applied to obtain the GTVp and GTVn segmentations on the test set. Soft (i.e., averaged label predictions for each voxel before softmax) and hard (i.e., averaged label predictions for each voxel after softmax) majority voting were applied. The whole pipeline is shown in Fig. 2.

$$DSCagg = \frac{2 \sum_i^N \sum_k \widehat{y}_{i,k} y_{i,k}}{\sum_i^N \sum_k (\widehat{y}_{i,k} + y_{i,k})} \tag{4}$$

Additionally, simple data augmentation was applied on-the-fly on each model run which consisted of: small random rotations and elastic deformations with a probability of 0.8 and random left-right flips with a probability of 0.5.

3   Results

The loss and learning rate employed to train the final selected models for each fold and structure can be found in Table 1. The training, validation, and test aggregated Dice coefficients can be found in Table 2. The organizers of the challenge only made available the average of the aggregated DSC for the GTVp and GTVn for the submitted approach (attempt3). The aggregated DSC for the GTVp and GTVn separately were made available for the model submitted outside of the ranking (attempt4). The difference between the two submission is that on attempt3 soft majority voting was applied, whereas in attempt4, hard majority voting was applied. Figure 3 shows the ground truth and predicted contours for four validation cases.

Table 1. Loss and learning rate (LR) of the selected models for each fold and each structure (GTVp and GTVn). DL: dice loss; TL: tversky loss; FTL: focal tversky loss

		Fold 1	Fold 2	Fold 3
GTVp	Loss	TL	DCE	DCE
	LR	0.001	0.001	0.001
GTVn	Loss	DCE	FTL	TL
	LR	0.001	0.001	0.001

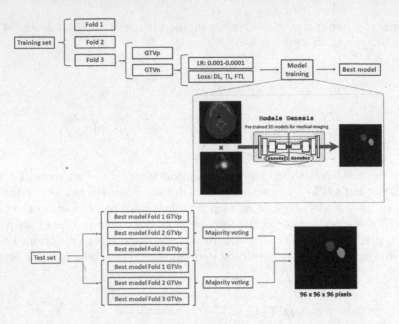

Fig. 2. Training and test pipelines of the proposed approach. GTVp: gross tumor volume of primary tumor; GTVn: gross tumor volume of affected lymph nodes; LR: learning rate; DL: Dice loss; TL: Tversky loss; FTL: focal Tversky loss. Models Genesis image taken from [6].

Table 2. Aggregated Dice coefficients for the training, validation and test sets for each fold and each structure (GTVp and GTVn). For the test set, only the average Dice coefficient is available.

		GTVp	GTVn	Average
Training	*Fold1*	0.720	0.701	0.711
	Fold2	0.720	0.718	0.719
	Fold3	0.772	0.719	0.746
Validation	*Fold1*	0.756	0.721	0.739
	Fold2	0.780	0.723	0.752
	Fold3	0.731	0.749	0.740
Test (attempt4)		0.720	0.667	0.690
Test (attempt3 HECKTOR ranking)				0.668

4   Discussion

In this study, we presented a transfer learning-based approach for automatic segmentation of the primary tumor and affected lymph nodes in fused PET/CT scans of oropharyngeal cancer patients. Transfer learning was performed from the publicly available Genesis Chest CT model [6], a 3D U-net pre-trained on

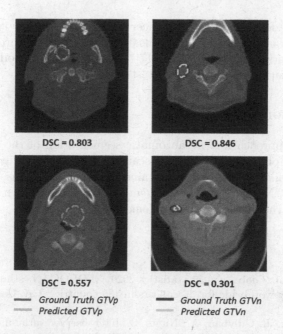

Fig. 3. Ground truth and predicted segmentations of the GTVp and GTVn for four patients in the validation test.

chest CT images. We opted for this approach as Zhou et al. showed a consistently superior performance in medical image segmentation and classification tasks of models which had been pre-trained on 3D medical data, compared to training from scratch or pre-training with natural images datasets, such as ImageNet. These findings have been also supported in others studies [12 14].

The results from the three-fold cross-validation within the training dataset showed promising Dice scores, whereas the model's performance on the test set was moderate. Attempt 4 showed comparable performance on the external test set for the segmentation of the GTVp, whereas the proposed approach failed to perform equally well on the GTVn segmentation task.

Different limitations were found in this study. Firstly, due to computational power constraints, we opted to study different binary segmentation models, instead of multi-class models. As a result, the output of a model designed to segment the GTVp and the output of a model for GTVn segmentation could overlap, which would not have been possible in the multi-class model. For the same reason and following the work of Saeed et al. presented in the HECKTOR challenge 2021 [10], we decided to employ a fused PET/CT input, instead of analyzing CT and PET separately by the network. This approach, even though it is simple, efficient and was proven by the authors to lead to the best performance, potentially discards useful information from both modalities. Another potential limitation is the inter-observer variability present within the dataset, which may hinder the performance of the model. Pavic et al. studied the inter-observer variability in tumor delineation in HN cancer among other cancer types and found that the mean [range] DSC of the delineations from three different expert radiation oncologists was of 0.72[0.21–0.89] [11], similar to our model performance for tumor delineation on the external test set.

5 Conclusion

The proposed approach for the automatic segmentation of the primary tumor and affected lymph nodes in PET/CT scans of OPC consisted in the fine-tuning of a publicly available 3D U-net pre-trained on chest CT scans, and performed moderately well. Nevertheless, better generalizability on unseen data should be achieved before the model can be implemented in the clinic.

References

1. Sung, H., et al.: Global cancer statistics 2020: GLOBOCAN estimates of incidence and mortality worldwide for 36 cancers in 185 countries. CA: Cancer J. Clin. **71**, 209–249 (2021)
2. van der Veen, J., Gulyban, A., Nuyts, S.: Interobserver variability in delineation of target volumes in head and neck cancer. Radiother. Oncol. **137**, 9–15 (2019)
3. La Macchia, M., et al.: Systematic evaluation of three different commercial software solutions for automatic segmentation for adaptive therapy in head-and-neck, prostate and pleural cancer. Radiat. Oncol. **7**, 160 (2012)
4. Oreiller, V., et al.: Head and neck tumor segmentation in PET/CT: the HECKTOR challenge. Med. Image Anal. **77**, 102336 (2022)
5. Andrearczyk, V., et al.: Overview of the HECKTOR challenge at MICCAI 2021: automatic head and neck tumor segmentation and outcome prediction in PET/CT images. In: Andrearczyk, V., Oreiller, V., Hatt, M., Depeursinge, A. (eds.) HECKTOR 2021. LNCS, vol. 13209, pp. 1–37. Springer, Cham (2022). https://doi.org/10.1007/978-3-030-98253-9_1
6. Zhou, Z., Sodha, V., Pang, J., Gotway, M.B., Liang, J.: Models genesis. Med. Image Anal. **67**, 101840 (2021)
7. Sudre, C.H., Li, W., Vercauteren, T., Ourselin, S., Jorge Cardoso, M.: Generalised dice overlap as a deep learning loss function for highly unbalanced segmentations. In: Cardoso, M.J., et al. (eds.) DLMIA/ML-CDS 2017. LNCS, vol. 10553, pp. 240–248. Springer, Cham (2017). https://doi.org/10.1007/978-3-319-67558-9_28
8. Salehi, S.S.M., Erdogmus, D., Gholipour, A.: Tversky loss function for image segmentation using 3D fully convolutional deep networks (2017). http://arxiv.org/abs/1706.05721
9. Abraham, N., Khan, N.M.: A Novel Focal Tversky loss function with improved Attention U-Net for lesion segmentation (2018). http://arxiv.org/abs/1810.07842
10. Saeed, N., Al Majzoub, R., Sobirov, I., Yaqub, M.: An ensemble approach for patient prognosis of head and neck tumor using multimodal data. In: Andrearczyk, V., Oreiller, V., Hatt, M., Depeursinge, A. (eds.) HECKTOR 2021. LNCS, vol. 13209, pp. 278–286. Springer, Cham (2022). https://doi.org/10.1007/978-3-030-98253-9_26
11. Pavic, M., et al.: Influence of inter-observer delineation variability on radiomics stability in different tumor sites. Acta Oncol. **57**, 1070–1074 (2018)

12. Alzubaidi, L., et al.: Novel transfer learning approach for medical imaging with limited labeled data. Cancers (Basel). **13**, 1590 (2021)

13. Matsoukas, C., Haslum, J.F., Sorkhei, M., Soderberg, M., Smith, K.: What makes transfer learning work for medical images: feature reuse & other factors. In: 2022 IEEE/CVF Conference on Computer Vision and Pattern Recognition (CVPR), New Orleans, LA, USA, pp. 9215–9224. IEEE (2022)

14. Alzubaidi, L., et al.: Towards a better understanding of transfer learning for medical imaging: a case study. Appl. Sci. **10**, 4523 (2020)

A U-Net Convolutional Neural Network with Multiclass Dice Loss for Automated Segmentation of Tumors and Lymph Nodes from Head and Neck Cancer PET/CT Images

Shadab Ahamed[1,2(✉)] ⓘ, Luke Polson[1,2], and Arman Rahmim[1,2,3] ⓘ

[1] Department of Physics and Astronomy, University of British Columbia, Vancouver, BC, Canada
shadab@phas.ubc.ca
[2] Department of Integrative Oncology, BC Cancer Research Institute, Vancouver, BC, Canada
[3] Department of Radiology, University of British Columbia, Vancouver, BC, Canada

Abstract. We implemented a 2D U-Net model with an ImageNet-pretrained ResNet50 encoder for performing segmentation of primary tumors (GTVp) and metastatic lymph nodes (GTVn) from PET/CT images provided by the HEad and neCK TumOR segmentation challenge (HECKTOR) 2022. We utilized a multiclass Dice Loss for model training which was minimized using the AMSGrad variant of the Adam algorithm optimizer. We trained our 2D models on the axial slices of the images in a 5-fold cross-validation setting and stacked the 2D predictions axially to obtain the predicted 3D segmentation masks. We obtained mean aggregate Dice similarity coefficients (mean DSC_{agg}) of 0.6865, 0.6689, 0.6768, 0.6792, and 0.6726 on the 5 validation sets respectively. The model with the best performance on the validation set (validation split 1) was chosen for evaluating segmentation masks on the test set for submission to the challenge. Our model achieved a mean $DSC_{agg} = 0.6345$ on the test set, with $DSC_{agg}(GTVp) = 0.6955$ and $DSC_{agg}(GTVn) = 0.5734$. The implementation can be found under our Github repository.

Keywords: Head and neck cancer · PET-CT · Segmentation · 2D U-Net · ResNet50 encoder · Multiclass dice loss

1 Introduction

Head and neck (H&N) cancer is the seventh most common cancer globally; there were an estimated 890 thousand new cases and 450 thousand deaths in 2018 [21]. Oropharyngeal cancer is the most prevalent type of malignant tumor in H&N cancer, accounting for about 32.2% of all H&N cancer patients [11]. Throughout cancer treatment, positron emission tomography and computed tomography (PET/CT) images are taken and used for tumor delineation; identification of

V. Andrearczyk et al. (Eds.): HECKTOR 2022, LNCS 13626, pp. 94–106, 2023.
https://doi.org/10.1007/978-3-031-27420-6_10

tumor volumes permits subsequent radiotherapy planning and surgical intervention [30]. At present, delineated tumors are qualitatively inferred from medical images by relevant experts, such as radiologists or nuclear medicine physicians. While manual delineation can be time-consuming, it is further prone to inconsistencies and errors attributed to the subjective experience of the physician [2]. It follows that automated segmentation of the regions of interest (ROIs) in H&N cancer is important for clinical decision-making and increasing patient throughput [3,26].

A number of semi-automated and fully automated methods have been developed over the years for tumor segmentation. A family of methods based on image thresholding at a relative of absolute image intensity values have made it to clinical practice [6,12]. However, thresholding typically results in healthy regions with high FDG uptake being selected, such as the cortex, bladder, and kidneys, which necessitates frequent manual intervention. More advanced methods of automated segmentation have been proposed that use traditional machine learning methods and various image features, such as contextual features [6,17]. Despite a large number of proposed methods, they have found little adoption so far in clinical PET/CT beyond individual research labs, due to several factors. First, PET imaging has a relatively lower resolution (~3–4 mm in clinical scanners) compared to other modalities and suffers from high noise levels due to the limited sensitivity of gamma detectors. Secondly, there is generally lower availability of PET images for method testing compared to other modalities; a related issue is the lack of standardized datasets for method evaluation [16]. Thirdly, there is a large variation in the shape and intensity of tumors, relating to a large heterogeneity in histology and genetics [7]; this renders segmentation of tumors a more challenging task due to the lack of robust anatomical shape priors.

In recent years, deep-learning techniques based on convolutional neural networks (CNNs) have achieved exceptional performance on several computer vision tasks for medical image analysis. A recent study in low-dose computed tomography for lung cancer screening found that a CNN architecture was able to outperform a sample of radiologists, with an 11% reduction in false positives and a 5% reduction in false negatives [5]. Another study on breast cancer screening found that a classifier obtained by averaging the probability of malignancy predicted by a radiologist and CNN was more accurate than either of the two separately [33]. There have also been recent reports of achieving human-level performance in image-based skin cancer detection [15] and cancer detection from X-ray mammography images [24], These improvements in medical image diagnosis could be sufficient to enable automated tumor segmentation in clinical practice, requiring minimal or no manual intervention, thus facilitating a routine collection and use of imaging biomarkers such as the total metabolic tumor volume, tumor dissemination, and other radiomic features [10,14,27,32].

Some of the advances in deep learning in medical image analysis may be attributed to the increased number of healthcare-based data challenge events, hosted by MICCAI, ISBI, SPIE Medical Imaging, and others, which attract participants from all over the world every year [REF]. An example of a noteworthy development is the nnU-Net architecture [20], which was developed and validated using the ten datasets provided by the Medical Segmentation Decathlon [4]. This work explores the predictive capabilities of a 2D U-Net neural network architecture in Task 1 of MICCAI HEad and neCK TumOR (HECKTOR) segmentation and outcome prediction challenge, 2022.

2   Materials and Methods

2.1   Data Description

For the challenge, we were provided with PET/CT images from patients with histologically proven oropharyngeal H&N cancer who underwent radiotherapy and/or chemotherapy treatment planning. These images came from FDG-PET and low-dose non-contrast-enhanced CT images (acquired with combined PET/CT scanners) of the H&N region from 9 different institutions, including four centers in Canada [31], two in Switzerland [8,9], two in France [23], and one in the USA [13]. A total of 524 PET/CT images were provided for training for Task 1 (segmentation of GTVp and GTVn) of the challenge. For the training cases, we were also provided with 3D ground truth (GT) segmentation masks, with labels 0 (for background), 1 (for GTVp), and 2 (for GTVn). A cohort of 329 PET/CT images (without GT masks) collected from three different centers were provided for testing. The training images were randomly split into training and validation sets (80:20%) in 5 different ways to set up 5-fold cross-validation during model training.

2.2   Image Preprocessing

The 3D CT and 3D GT images were first resampled to their respective 3D PET coordinates using bilinear and nearest neighbor interpolation, respectively [34]. The CT images were median filtered using a window size of $5 \times 5 \times 5$ to remove noise. The CT HU values were clipped between $[-200, 300]$ HU since this was the range of HU values for the GTVp and GTVn in the resampled CT images in the training set. Similarly, all the PET SUV values were clipped between $[0, 50]$ SUV, since the maximum SUV values inside the GTVp and GTVn regions in all the PET images in the training set was SUV ≈ 45. The clipped values of CT and PET were subsequently normalized using $CT_{normalized} = (CT + 200)/(300 + 200)$ and $PET_{normalized} = PET/50$.

After resampling, clipping, and normalization, the 3D images were sliced into 2D axial slices. From a total of 524 cases in the HECKTOR 2022 training set, we obtained 110096 axial slices, which were split into 5 training and validation sets as per the 5 folds (for cross-validation) generated based on Patient-ID, as shown in Table 1.

Table 1. Number of 2D axial slices in the 5 folds of training and validation set.

Fold	Number of training slices	Number of validation slices
1	88749	21347
2	87380	22716
3	88249	21847
4	87115	22981
5	88510	21586

2.3   Network Architecture

A 2D U-Net with an ImageNet-pretrained residual neural network (ResNet50) encoder [18,19,29] was implemented (as shown in Fig. 1) for performing the segmentation of GTVp (label = 1) and GTVn (label = 2) on the axial slices of 3D images. This pretrained network architecture was adapted from the segmentation-models-pytorch repository [19]. The input to the network was created by stacking two same PET axial slices followed by the two corresponding CT slices, leading to a 4-channel image [PET slice, PET slice, CT slice, CT slice] (the architectures implemented in [19] support customized number of input/output channels for pretrained encoders). All the axial slices were resized to 128 × 128 pixels before giving them as input to the network. The output of the model was a one-hot encoded 3-channel image of size 128 × 128 with the first channel providing prediction of the background pixels, the second for GTVp, and the third for GTVn.

Additionally, we used the 2D models with encoders that were pretrained on ImageNet dataset to utilize the power of transfer learning [22,35]. The models were trained for 100 epochs (as explained in Sect. 2.4). To prevent overfitting while training, the pretrained encoder weights were initially frozen for the first 75 epochs, and later the whole network was fine-tuned for the remaining 25 epochs.

The above-mentioned model was the best-performing model among the several model architectures that were tried for this challenge. Several 2D networks with different pretrained encoder backbones (like ResNet18, ResNet34, ResNet50, ResNet101, ResNet152, etc.) and different number of input channels (2: [PET slice, CT slice], 3: [PET slice, PET slice, CT slice], 3: [PET slice, PET slice, PET slice], 4: [PET slice, PET slice, CT slice, CT slice], etc.) were tried and the model with ImageNet-pretrained ResNet50 encoder with 4 input channels [PET slice, PET slice, CT slice, CT slice] performed the best on the internal validation sets. Other than the 2D networks, we also tried 3D UNet and Attention-UNet models as implemented in the package MONAI. In our experiments, the performances of the 3D models were always lower than those of 2D models (the best 3D model achieved a mean $DSC_{agg} = 0.5982$). This could mainly be attributed to a lack of extensive hyperparameter search for the case of 3D models. The 2D models also might have performed better without extensive hyperparameter tuning since they were trained on axial slices of the 3D images (rather than on the 3D images), which led to them getting trained on a larger number of images.

2.4 Model Training

The model was trained using the following multiclass Dice (MCD) loss [36] function, calculated on axial slices:

$$\mathcal{L}_{\mathrm{MCD}} = 1 - 2\frac{\sum_{k \in C}\sum_{i \in N} y_i^{(k)}\hat{y}_i^{(k)}}{\sum_{k \in C}\sum_{i \in N}(y_i^{(k)} + \hat{y}_i^{(k)}) + \varepsilon} \tag{1}$$

where $y_i^{(k)}$ and $\hat{y}_i^{(k)}$ are the GT and predicted segmentation masks respectively for image i and class k, $C = \{0, 1, 2\}$ (0 for background, 1 for GTVp, and 2 for GTVn) denotes the indices of different classes in the segmentation masks, N is the total number of images in a batch during training, and $\varepsilon = 10^{-7}$ is a small constant to prevent division by 0 errors while computing the loss. We used $N = 256$ for all our training procedures.

The L_{MCD} was minimized using the AMSGrad variant of the Adam algorithm optimizer [28] with an initial learning rate $= 3 \times 10^{-3}$, and a weight-decay $= 3 \times 10^{-5}$. The learning rate scheduler ReduceLROnPlateau (implemented in Pytorch [1]) was employed with a `factor=0.1`, `patience=10`, and `threshold=0.001`. This reduced the learning rate by a factor of 0.1, after waiting for 10 epochs for the validation loss to decrease whenever the loss plateaued (change `threshold=0.001`). The models were trained/validated on the five folds of training/validation sets for 100 epochs and the model with the smallest loss on the validation set was chosen for evaluation from each of the five folds. Finally, out of the 5 models trained on each of the 5 folds, the model with the best mean $\mathrm{DSC}_{\mathrm{agg}}$ between GTVp and GTVn on their respective validation sets was chosen as the best model for further analysis.

2.5 Image Postprocessing

The one-hot encoded 3-channel model output of size 128×128 was first converted to a 1-channel output of labels (containing 0, 1, or 2) and was then resized back to its original PET axial image size. The predicted axial slices belonging to the same patient were then stacked together to generate the final predicted 3D segmentation mask. This mask was resampled to the original CT coordinates using nearest neighbor interpolation for further evaluations of performance metrics.

3   Results

3.1   Segmentation Performance

We used the 5 best models obtained from each fold of cross-validation to compute performance measures on the respective validation sets. We evaluated the

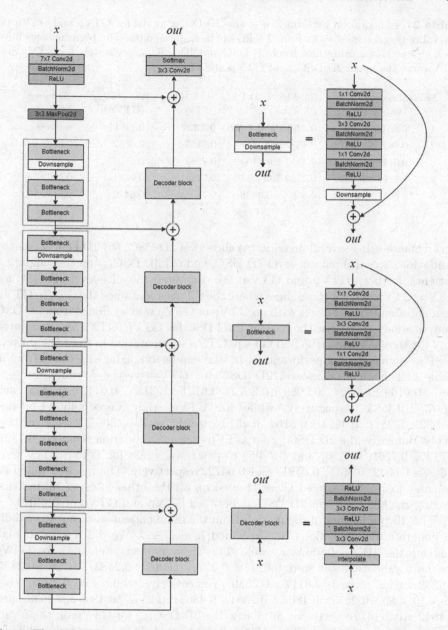

Fig. 1. The 2D UNet network architecture used in this work (shown on the left). It consists of a ResNet50 encoder (for feature extraction) and decoder branches. The ResNet50 encoder is built using bottleneck residual modules [18], which are variants of residual modules that utilize 1×1 convolutions to create a bottleneck. The bottleneck is useful since it reduces the number of matrix multiplications, with the idea to make residual blocks as thin as possible thereby increasing network depth and reducing the number of parameters. The various building blocks of the network are described on the right.

Table 2. Segmentation performance of the 2D U-Net model for GTVp and GTVn on the validation set axial slices from 5 folds of the training data. The performances have been characterized using slice-level 2D DSC and 2D DSC$_{agg}$ separately for GTVp and GTVn and the mean 2D DSC$_{agg}$ of GTVp and GTVn.

Fold	Slice-level 2D DSC (GTVp)	Slice-level 2D DSC (GTVn)	2D DSC$_{agg}$ (GTVp)	2D DSC$_{agg}$ (GTVn)	2D mean DSC$_{agg}$
1	**0.6169 ± 0.3224**	**0.4560 ± 0.3428**	**0.7327**	**0.6644**	**0.6986**
2	0.6058 ± 0.3672	0.4532 ± 0.2877	0.7135	0.6532	0.6834
3	0.6141 ± 0.3185	0.4553 ± 0.3164	0.7261	0.6578	0.6920
4	0.6113 ± 0.3352	0.4512 ± 0.3217	0.7286	0.6591	0.6939
5	0.6077 ± 0.3583	0.4509 ± 0.3432	0.7205	0.6472	0.6839

performance using several metrics: (a) slice-level 2D DSC, (b) 2D DSC$_{agg}$ on the validation slices, (c) patient-level 3D DSC, and (d) 3D DSC$_{agg}$ on the validation patients, for both GTVp and GTVn. We computed the slice-level 2D DSC for GTVp/GTVn only for those slices where the GT slice contained the GTVp/GTVn segmentation (the GT slices with no GTVp or GTVn were excluded from 2D DSC computation). Similarly, the patient-level DSC for GTVp/GTVn was computed only for those patients who had GTVp/GTVn segmentation in their GT masks.

The segmentation performance of the model on axial slices is given in Table 2. The mean slice-level 2D DSC for GTVp on the 5 validation sets were 0.6169 ± 0.3224, 0.6058 ± 0.3672, 0.6141 ± 0.3185, 0.6113 ± 0.3352, and 0.6077 ± 0.3583, respectively, while for GTVn, they were 0.4560 ± 0.3428, 0.4532 ± 0.2877, 0.4553 ± 0.3164, 0.4512 ± 0.3217, and 0.4509 ± 0.3432, respectively. Similarly, the 2D DSC$_{agg}$ for GTVp on the 5 validation sets were 0.7327, 0.7135, 0.7261, 0.7286, and 0.7205, respectively, while for GTVn, they were 0.6644, 0.6532, 0.6578, 0.6591, and 0.6472, respectively. The model trained on the first fold outperformed those trained on all the other folds on all metrics, scoring the highest mean 2D DSC$_{agg}$ between GTVp and GTVn of **0.6986**.

The 2D predictions of the models from the same patient were stacked axially to generate 3D predictions, as explained in Sect. 2.5. We did a similar analysis for the 3D predictions as well. The mean patient-level DSC for GTVp on the 5 validation sets were 0.6247 ± 0.2221, 0.6092 ± 0.2483, 0.6228 ± 0.2855, 0.6232 ± 0.2372, and 0.6117 ± 0.2530, respectively, while for GTVn, they were 0.4959 ± 0.2636, 0.4801 ± 0.2581, 0.4869 ± 0.2452, 0.4865 ± 0.2533, and 0.4842 ± 0.2714, respectively. Similarly, the 3D DSC$_{agg}$ for GTVp on the 5 validation sets were 0.7350, 0.7172, 0.7305, 0.7311, and 0.7249 respectively, while for GTVn, they were 0.6379, 0.6205, 0.6231, 0.6273, and 0.6202 respectively. Again, the performance was the best on the first fold with a mean 3D DSC$_{agg}$ = **0.6865**, with the DSC$_{agg}$(GTVp) = 0.7350 and DSC$_{agg}$(GTVn) = 0.6379. Some examples comparing the 3D GT and 3D predicted masks for GTVp and GTVn for some of the patients from the validation set are shown in Fig. 2 (Table 3).

Table 3. Evaluation of 3D predictions obtained by axial stacking of 2D prediction for each patient. The 3D predictions were resampled to original CT coordinates and the patient-level DSC and DSC_{agg} were computed for both GTVp and GTVn, followed by the mean DSC_{agg} of GTVp and GTVn on the 5 validation sets.

Fold	Patient-level 3D DSC (GTVp)	Patient-level 3D DSC (GTVn)	3D DSC_{agg} (GTVp)	3D DSC_{agg} (GTVn)	3D mean DSC_{agg}
1	**0.6247 ± 0.2221**	**0.4959 ± 0.2636**	**0.7350**	**0.6379**	**0.6865**
2	0.6092 ± 0.2483	0.4801 ± 0.2581	0.7172	0.6205	0.6689
3	0.6228 ± 0.2855	0.4869 ± 0.2452	0.7305	0.6231	0.6768
4	0.6232 ± 0.2372	0.4865 ± 0.2533	0.7311	0.6273	0.6792
5	0.6117 ± 0.2530	0.4842 ± 0.2714	0.7249	0.6202	0.6726

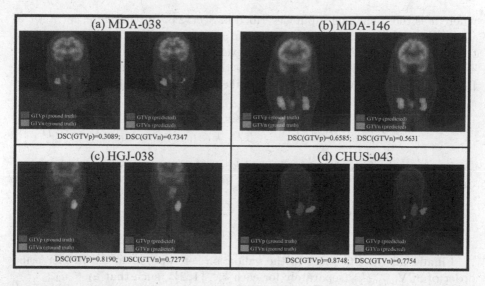

Fig. 2. Some examples showing the patient-level 3D DSC(GTVp) and DSC(GTVn) between the ground truth and predicted segmentation masks of 4 patients, namely (a) MDA-038, (b) MDA-146, (c) HGJ-038, and (d) CHUS-043, in the coronal view. The GTVp and GTVn segmentation are shown in red and green, respectively. (Color figure online)

3.2 Model Performance Analysis on Slices Containing GTVp or GTVn

We further analyzed the performance of the 2D U-Net model by understanding its prediction on the slices containing GTVp or GTVn labels (foreground (FG) slices). For this analysis, we only used the model trained on fold 1 and analyzed its prediction on the respective validation set. Firstly, for each GT slice of size $M \times M$ with class $k (\in \{1, 2\})$, we define the area of the ROI of label k as,

$$A(k) = \frac{1}{k} \sum_{\text{GT}(i,j)=k} \text{GT}(i,j), \text{ for } k \in \{1,2\} \tag{2}$$

where (i,j) is a pixel location on the GT slice. Finally, the normalized area of the ROI k on a slice s can be defined as

$$L_s(k) = \frac{A_s(k)}{M^2} = \frac{1}{M^2 k} \sum_{\text{GT}(i,j)=k} \text{GT}(i,j), \text{ for } k \in \{1,2\} \tag{3}$$

In addition, for a slice s_i, the model ($\mathcal{M}$) prediction can be defined as,

$$\text{Pr}_{s_i} = \mathcal{M}(x_{s_i}) \tag{4}$$

where x is the 4-channel input described in Sect. 2.3. Furthermore, for a set of slices $\{s_1, s_2, ..., s_n\}$ and their respective predictions $\{\text{Pr}_{s_1}, \text{Pr}_{s_2}, ..., \text{Pr}_{s_n}\}$, the true positive rate (TPR) at the pixel-level for class k is given by $\text{TPR}(k) = \text{TP}(k)/\text{P}(k)$, where $\text{TP}(k)$ is the true-positive segmented pixels for label k, and $\text{P}(k)$ is the positive pixels for label k in the respective GT slices.

$$\text{TPR}(k) = \frac{\sum_{t=1}^{n} \text{Pr}_{s_t}(k) \cap \text{GT}_{s_t}(k)}{\sum_{t=1}^{n} A_{s_t}(k)} = \frac{\sum_{t=1}^{n} [\text{Pr}_{s_t}(i,j) = k] \cap [\text{GT}_{s_t}(i,j) = k]}{\frac{1}{k} \sum_{t=1}^{n} \sum_{\text{GT}_{s_t}(i,j)=k} \text{GT}_{s_t}(i,j)} \tag{5}$$

Furthermore, for a slice s with label k, we also defined a quantity that gives the measure of PET SUV_{mean} inside the ROI k, as

$$\text{SUV}_{\text{mean},s}(k) = \frac{1}{A_s(k)} \sum_{\text{GT}_s(i,j)=k} PET_s(i,j) \tag{6}$$

Finally, all the FG slices in the validation set, were arranged in increasing order of $\text{SUV}_{\text{mean}}(k)$ (separately for each $k \in \{1,2\}$), such that $u_1 \leq u_2 \leq ... \leq u_{n-1} \leq u_n$, where u_i is the tumor SUV_{mean} value of the FG slice with the ith largest SUV_{mean} and n is the number of slices in the validation set. Within the validation set, different subsets of slices were chosen such that subset i consisted of all the slices with tumor $\text{SUV}_{\text{mean}} \geq u_i$. Hence, the subset $i = 1$ consisted of all n slices of the validation set, while subset $i = n$ consisted of only the slice with tumor $\text{SUV}_{\text{mean}} = u_n$. The $\text{TPR}(k)$ was computed for $k \in \{1,2\}$ for each of these subsets and plotted as a function of u_i for $1 \leq i \leq floor(0.95n)$, as shown in Fig. 3(a) and (c) for $k = 1$ (GTVp) and $k = 2$ (GTVn), respectively. Hence, by this analysis, we demonstrate that the TPR increases for both GTVp and GTVn for a set of slices with higher PET SUV_{mean} inside the GTVp or GTVn regions respectively. Hence, the model is more likely to segment an ROI correctly, if the SUV_{mean} inside the ROI is '*larger*'.

A similar analysis as above was conducted for the normalized ROI size metric, as defined in Eq. 3 for both GTVp and GTVn. From Fig. 3(b), we can conclude that TPR increases for a set of slices with '*larger*' average GTVp sizes on the axial slices. A similar conclusion cannot be drawn for the GTVn (Fig. 3(d)), since

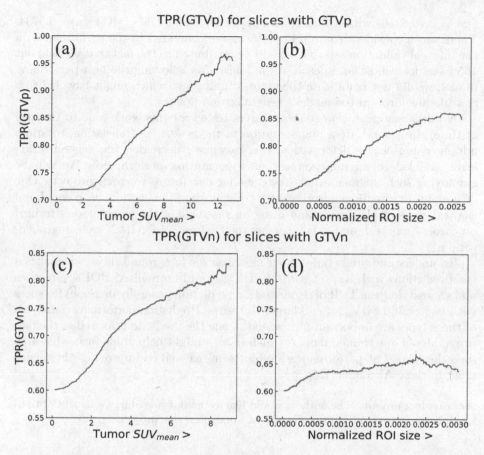

Fig. 3. Model prediction analysis on the foreground axial slices in the validation set. These plots demonstrate that it was easier for the model to correctly segment an ROI with a higher PET SUV_{mean} (a, c) or a larger size (b, d), though the increase in TPR from smaller to larger sized GTVn on the axial slices (d) is much smaller as compared to GTVp (b).

the TPR only increases slightly as the average GTVn size increases in the subset. We speculate that this could be attributed to the fact that the larger GTVn's had lower SUV_{mean} on an average which led to a poor model segmentation performance for them, although further analysis is required to make a concrete inference.

4 Conclusion and Discussion

We trained a 2D U-Net model on the axial slices of PET/CT images to predict multiclass 2D segmentation masks, which were stacked to generate patient-level 3D segmentation. On the challenge test set, we obtained a mean

$DSC_{agg} = 0.6345$, with $DSC_{agg}(GTVp) = 0.6955$ and $DSC_{agg}(GTVn) = 0.5734$. There was a considerable drop in performance compared to the performance on the internal validation sets. This could be attributed to the fact that we held out 105 cases for validation in each of the folds; after selecting the best-performing model, we did not train it on the entire training set, which might have led to a considerable drop in the model's generalization power.

No data augmentation strategies were tested for this work (due to the lack of time). In general, data augmentation methods such as translation, rotation, adding noise, elastic deformation, etc. may have decreased the generalization error and led to an improvement in segmentation performance. We will be employing such augmentation strategies for our future research projects. One of our future directions also includes training the 2D networks on coronal and sagittal slices of 3D images and using an ensemble of the three models trained on coronal, sagittal, and axial slices as the final model for H&N cancer segmentation.

To understand predictions of our model on the foreground slices, we analyzed the predictions with respect to ROI SUV_{mean} and normalized ROI size metrics and showed that the TPR of segmentation prediction generally increases for slices with higher ROI SUV_{mean} and larger ROI sizes. Prediction performance analyses of these types are important for understanding the predictions as well as the failure modes of our trained models which in turn might help guide future research direction aimed at performance improvement, as well as improving physicians' trust in these AI models [25].

Acknowledgement. The authors would like to thank Isaac Shiri for insightful technical discussions throughout the course of working on this project.

References

1. ReduceLROnPlateau. https://pytorch.org/docs/stable/generated/torch.optim.lr_scheduler.ReduceLROnPlateau.html
2. Ahamed, S., et al.: A cascaded deep network for automated tumor detection and segmentation in clinical PET imaging of diffuse large B-cell lymphoma. In: Colliot, O., Išgum, I. (eds.) Medical Imaging 2022: Image Processing, vol. 12032, p. 120323M. International Society for Optics and Photonics, SPIE (2022). https://doi.org/10.1117/12.2612684
3. Andrearczyk, V., et al.: Overview of the HECKTOR challenge at MICCAI 2021: automatic head and neck tumor segmentation and outcome prediction in PET/CT images (2022). https://doi.org/10.48550/ARXIV.2201.04138, https://arxiv.org/abs/2201.04138
4. Antonelli, M., et al.: The medical segmentation Decathlon (2021). https://doi.org/10.48550/ARXIV.2106.05735, https://arxiv.org/abs/2106.05735
5. Ardila, D., Kiraly, A., Bharadwaj, S., et al.: End-to-end lung cancer screening with three-dimensional deep learning on low-dose chest computed tomography. Nat. Med. **25**(4), 954–961 (2019). https://doi.org/10.1038/s41591-019-0447-x. Accessed 07 Sep 2022

6. Bi, L., Kim, J., Feng, D., Fulham, M.: Multi-stage thresholded region classification for whole-body PET-CT lymphoma studies. Med. Image Comput. Comput. Assist. Interv. **17**(Pt 1), 569–576 (2014)

7. Blanc-Durand, P., Van Der Gucht, A., Schaefer, N., Itti, E., Prior, J.O.: Automatic lesion detection and segmentation of 18F-FET PET in gliomas: a full 3D U-Net convolutional neural network study. PLoS ONE **13**(4), e0195798 (2018)

8. Bogowicz, M., et al.: Comparison of PET and CT radiomics for prediction of local tumor control in head and neck squamous cell carcinoma. Acta Oncol. **56**(11), 1531–1536 (2017)

9. Castelli, J., et al.: A PET-based nomogram for oropharyngeal cancers. Eur. J. Cancer **75**, 222–230 (2017)

10. Cottereau, A.S., et al.: New approaches in characterization of lesions dissemination in DLBCL patients on baseline PET/CT. Cancers **13**(16), 3998 (2021). https://doi.org/10.3390/cancers13163998, https://europepmc.org/articles/PMC8392801

11. DiGiulio, S.: Oropharyngeal cancer now most common head & neck cancer. Oncol. Times **36**(22) (2014). https://journals.lww.com/oncology-times/Fulltext/2014/11250/Oropharyngeal_Cancer_Now_Most_Common_Head__Neck.26.aspx

12. Driessen, J., et al.: Baseline metabolic tumor volume in 18FDG-PET-CT scans in classical Hodgkin lymphoma using semi-automatic segmentation. Blood **134**, 4049 (2019). https://doi.org/10.1182/blood-2019-125495, https://www.sciencedirect.com/science/article/pii/S0006497118619779

13. Ger, R.B., et al.: Radiomics features of the primary tumor fail to improve prediction of overall survival in large cohorts of CT- and PET-imaged head and neck cancer patients. PLoS ONE **14**(9), e0222509 (2019)

14. Guo, B., Tan, X., Ke, Q., Cen, H.: Prognostic value of baseline metabolic tumor volume and total lesion glycolysis in patients with lymphoma: a meta-analysis. PLoS ONE **14**(1), e0210224 (2019)

15. Haenssle, H.A., et al.: Man against machine: diagnostic performance of a deep learning convolutional neural network for dermoscopic melanoma recognition in comparison to 58 dermatologists. Ann. Oncol. **29**(8), 1836–1842 (2018)

16. Hatt, M., et al.: The first MICCAI challenge on PET tumor segmentation. Med. Image Anal. **44**, 177–195 (2018)

17. Hatt, M., et al.: Classification and evaluation strategies of auto-segmentation approaches for PET: report of AAPM task group no. 211. Med. Phys. **44**(6), e1–e42 (2017)

18. He, K., Zhang, X., Ren, S., Sun, J.: Deep residual learning for image recognition (2015). https://doi.org/10.48550/ARXIV.1512.03385, https://arxiv.org/abs/1512.03385

19. Iakubovskii, P.: Segmentation models PyTorch (2019). https://github.com/qubvel/

20. Isensee, F., Jaeger, P., Kohl, S., et al.: nnU-Net: a self-configuring method for deep learning-based biomedical image segmentation. Nat. Methods **18**, 203–211 (2021)

21. Johnson, D.E., Burtness, B., Leemans, C.R., Lui, V.W.Y., Bauman, J.E., Grandis, J.R.: Head and neck squamous cell carcinoma. Nat. Rev. Dis. Primers. **6**(1), 92 (2020)

22. Kim, H.E., Cosa-Linan, A., Santhanam, N., Jannesari, M., Maros, M.E., Ganslandt, T.: Transfer learning for medical image classification: a literature review. BMC Med. Imaging **22**(1), 69 (2022). https://doi.org/10.1186/s12880-022-00793-7

23. Legot, F., et al.: Use of baseline 18F-FDG PET scan to identify initial sub-volumes with local failure after concomitant radio-chemotherapy in head and neck cancer. Oncotarget **9**(31), 21811–21819 (2018)
24. McKinney, S.M., et al.: International evaluation of an AI system for breast cancer screening. Nature **577**(7788), 89–94 (2020)
25. Nickel, P.J.: Trust in medical artificial intelligence: a discretionary account. Ethics Inf. Technol. **24**(1), 1–10 (2022). https://doi.org/10.1007/s10676-022-09630-5
26. Oreiller, V., et al.: Head and neck tumor segmentation in PET/CT: the HECKTOR challenge. Med. Image Anal. **77**, 102336 (2022). https://doi.org/10.1016/j.media.2021.102336, https://www.sciencedirect.com/science/article/pii/S1361841521003819
27. Orlhac, F., Nioche, C., Klyuzhin, I., Rahmim, A., Buvat, I.: Radiomics in PET imaging: a practical guide for newcomers. PET Clin. **16**(4), 597–612 (2021)
28. Reddi, S.J., Kale, S., Kumar, S.: On the convergence of Adam and beyond. In: International Conference on Learning Representations (2018). https://openreview.net/forum?id=ryQu7f-RZ
29. Ronneberger, O., Fischer, P., Brox, T.: U-net: convolutional networks for biomedical image segmentation (2015). https://doi.org/10.48550/ARXIV.1505.04597, https://arxiv.org/abs/1505.04597
30. Slattery, A.: Validating an image segmentation program devised for staging lymphoma. Australas. Phys. Eng. Sci. Med. **40**(4), 799–809 (2017)
31. Vallières, M., et al.: Radiomics strategies for risk assessment of tumour failure in head-and-neck cancer. Sci. Rep. **7**(1), 10117 (2017)
32. Vercellino, L., et al.: High total metabolic tumor volume at baseline predicts survival independent of response to therapy. Blood **135**(16), 1396–1405 (2020)
33. Wu, N., et al.: Deep neural networks improve radiologists' performance in breast cancer screening. IEEE Trans. Med. Imaging **39**(4), 1184–1194 (2020). https://doi.org/10.1109/TMI.2019.2945514
34. Yaniv, Z., Lowekamp, B.C., Johnson, H.J., Beare, R.: SimpleITK image-analysis notebooks: a collaborative environment for education and reproducible Research. J. Digit. Imaging **31**(3), 290–303 (2017). https://doi.org/10.1007/s10278-017-0037-8
35. Yosinski, J., Clune, J., Bengio, Y., Lipson, H.: How transferable are features in deep neural networks? (2014)
36. Zhou, T., Ruan, S., Canu, S.: A review: deep learning for medical image segmentation using multi-modality fusion. Array **3–4**, 100004 (2019). https://doi.org/10.1016/j.array.2019.100004

Multi-scale Fusion Methodologies for Head and Neck Tumor Segmentation

Abhishek Srivastava[1]($\boxtimes$) (iD), Debesh Jha[1] (iD), Bulent Aydogan[2] (iD), Mohamed E. Abazeed[3] (iD), and Ulas Bagci[1] (iD)

[1] Machine and Hybrid Intelligence Lab, Department of Radiology, Northwestern University, Chicago, USA
abhishek.srivastava@northwestern.edu
[2] Department of Radiation Oncology, University of Chicago, Chicago, IL, USA
[3] Department of Radiation Oncology, Northwestern University, Chicago, IL, USA

Abstract. Head and Neck (H&N) organ-at-risk (OAR) and tumor segmentations are an essential component of radiation therapy planning. The varying anatomic locations and dimensions of H&N nodal Gross Tumor Volumes (GTVn) and H&N primary gross tumor volume (GTVp) are difficult to obtain due to lack of accurate and reliable delineation methods. The downstream effect of incorrect segmentation can result in unnecessary irradiation of normal organs. Towards a fully automated radiation therapy planning algorithm, we explore the efficacy of multi-scale fusion based deep learning architectures for accurately segmenting H&N tumors from medical scans. Team Name: M&H_lab_NU.

Keywords: Tumor segmentation · Head and neck · Multi-scale fusion

1 Introduction

Optimizations in radiation treatment plans for Head and Neck (H&N) tumors have seen significant advancements in recent years. Quantitative imaging biomarkers obtained from medical scans have shown promise in modelling disease characteristics and treatment outcomes [1,8]. A prerequisite to radiation therapy (RT) is an accurate delineation of (H&N) tumors to obtain H&N nodal gross tumor volumes (GTVn) and H&N primary gross tumor volume (GTVp) from volumetric medical scans. Manual annotation of the region of interest requires significant content expertise and is both laborious and time-consuming, although being the gold standard. Instead, automated segmentation systems can swiftly provide segmentation maps of the region of interest and, consequently, improve patient care on a large scale. Since tumor size can vary, and the nature of the problem constitutes itself at varying scales, conventional deep learning algorithms provide only sub-optimal solutions for this problem. Recently, multi-scale fusion methodologies have shown great capacity in generating precise segmentation maps [4,11–14] when the object of interest exists in various different scales. Such methodologies have established their efficacy in the segmentation of 2-D

V. Andrearczyk et al. (Eds.): HECKTOR 2022, LNCS 13626, pp. 107–113, 2023.
https://doi.org/10.1007/978-3-031-27420-6_11

medical images. The repeated fusion of multi-scale features generates diverse and robust features and allows a more generalizable model [11], capable of modelling the varying size of the region of interest. We study the performance of such multi-scale fusion-based methodologies to obtain GTVn and GTVp from FluoroDeoxyGlucose (FDG)-Positron Emission Tomography (PET) and Computed Tomography (CT) scans. As participants in the HECKTOR 2022 challenge [2,9], we used the PET/CT images, GTVn masks, and GTVp masks released by the challenge organizers to train our two algorithms, named OARFocalFuseNet and 3D-MSF. We perform additional experiments with SwinUNETR [5], to compare the efficiency of self-attention mechanisms by Transformers [3,6,7] with multi-scale fusion techniques. The organization of the rest of the paper is as follows. Section 2 provides a brief description of all the methods used for our experiments. Section 3 provides the experiment and implementation details. Section 4 discusses the results obtained by OARFocalFuseNet [13], 3D-MSF [13], and SwinUNETR [5]. Finally, we conclude our paper in Sect. 5.

2   Method

In this section, we discuss the three different deep-learning methodologies used in our experiments: OARFocalFuseNet [13], 3D-MSF [13], and SwinUNETR [5].

2.1   Submission 1: OARFocalFuseNet

Let C and P be the input CT and PET scan where $C\&P \in \mathbb{R}^{W \times H \times Z}$. Here, W, H, and Z denote width, height, and length (number of slices), respectively. We concatenate C and P along the channel axis to form X before feeding it into the encoder. The encoder blocks employ convolutional layers and pooling layers to extract features for a particular resolution scale and then downscale the resolution by a factor of 2. Let $[X_1, X_2, X_3, X_4]$ be the sets of feature maps extracted by encoder blocks, each with a distinct resolution scale. Hereafter, a linear layer is used to transform the feature space X_a into $F_{a,0}$, where a denotes the resolution scale.

Multi-scale feature fusion is then performed by fusing multi-scale resolution features across all resolution streams (Fig. 1(b)). A combination of strided depthwise convolution and pooling layers is used to downscale the spatial dimensions of features being transmitted from higher to lower-resolution streams. Similarly, a combination of strided depth-wise deconvolution and bicubic interpolation layers are used to upscale the spatial dimensions of features being transmitted from lower to higher-resolution streams.

$$F_{a,l} = GeLU(DC_{3\times3}(Conv_{1\times1}(F_{a,l-1}, F_{b,l-1}, F_{c,l-1},$$
$$F_{d,l-1}))) \{b,c,d\} \neq a, \{a,b,c,d\} \in \{1,2,3,4\}. \quad (1)$$

Here, DC and $Conv$ represents a depth-wise convolutional layer and a standard convolutional layer, respectively. Additionally, l denotes the multi-scale focal

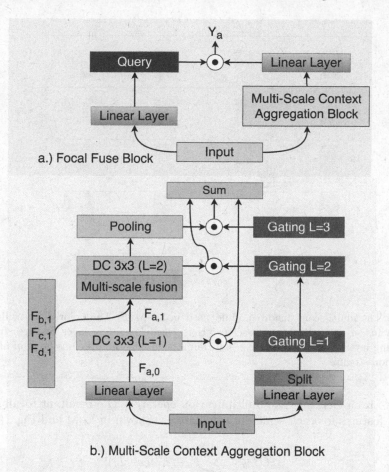

Fig. 1. The OARFocalFuseNet components. **a)** The Focal Fuse block, which aggregates multi-scale global-local context **b)** The Multi-Scale Context Aggregation block, which gathers multi-scale features and performs depthwise convolutions to gather features with diverse context ranges and performs spatial and channel-wise gating to prune irrelevant features.

level, with the total number of focal levels being N. Moreover, a linear layer is utilized for pruning extraneous features (see Eq. 2).

$$G_{a,l} = Linear(F_{a,0}). \tag{2}$$

where $G_a \in R^{W \times H \times Z \times N+1}$. The multi-scale focal modulator is calculated by adding the context information accumulated by each multi-scale focal layer (see Eq. 3 and Fig. 1(b)).

$$F_a = \sum_{l=1}^{N+1} F_{a,l} \odot G_{a,l}. \tag{3}$$

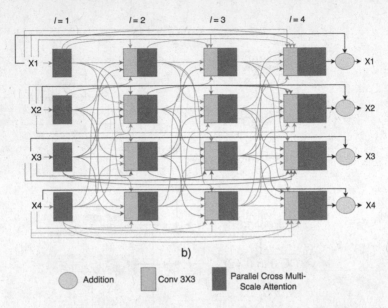

Fig. 2. The multi-scale fusion module used in 3D-MSF. The colored lines illustrate feature fusion at multiple cross-scales. Here, each layer l receives features from all preceding layers within the same resolution stream and the previous layer of all other resolution streams.

Here, $\odot$ is an element-wise multiplication operator. The resultant focally modulated features for each scale are calculated as shown in Eq. 4 and Fig. 1(a).

$$MOD_a = \sum_{a=1}^{4} F_a \odot Linear(I_a). \tag{4}$$

2.2   Submission 2: 3D Multi-scale Fusion Network

The 3D Multi-scale Fusion Network (3D-MSF) uses densely connected blocks to perform multi-scale feature fusion. Initially, an encoder identical to the one used by OARFocalFuseNet is used for feature extraction. Each set of feature maps with a distinct resolution has its own resolution stream, which comprises a densely connected block. In this block, each layer receives inputs from all preceding layers in the same resolution scale and the previous layer from all other resolution scales (see Eq. 5 and Fig. 2).

$$X_{a,l} = DepthConv_{3\times3}(X_{a,0} \oplus \cdots X_{a,l-1} \oplus X_{b,l-1} \oplus X_{c,l-1} \\ \oplus X_{d,l-1}), \{b,c,d\} \neq a, \{a,b,c,d\} \in \{1,2,3,4\}, \tag{5}$$

where l denotes the layer inside the dense blocks.

2.3 Submission 3: SwinUNETR

SwinUNETR follows the same architectural design as a standard UNet [10]. SwinUNETR consists of an encoder, bottleneck, decoder, and skip connections. The basic unit of SwinUNETR is the Swin Transformer block. The input is first to split into non-overlapping patches before being projected to another feature space using a linear layer. Subsequently, these patches pass through the patch merging and Swin Transformer blocks to extract features. Then the decoder uses Swin Transformer blocks and patch-expanding layers to upscale the features obtained from the encoder. Additionally, the features obtained from the decoder blocks are fused with the corresponding encoder features via skip-connections. Lastly, the last patch expanding layer is used to perform 4× up-sampling to restore the resolution of the feature maps to the input resolution (W × H × Z). The Swin Transformer block combines the window-based multi-head self-attention (W-MSA) [7] module and the shifted window-based multi-head self-attention (SW-MSA) [7] module in succession before performing the self-attention operation.

3 Experiments

3.1 Data Pre-processing and Data Augmentation

We use the maximum of the CT/PET origin and the minimum of the CT/PET size to crop the input CT and PET volumes. Hereafter, they both are re-sampled and set to have the same origin, direction and size. Next, we perform the standard practice of clipping all CT values greater than 300 and less than −300 before performing min-max scaling. The PET volumes are also normalized using min-max scaling before being concatenated with CT volumes along the channel axis. Our data augmentation scheme involves random cropping, random Affine transformation, random 3D elastic transformation, and random Gaussian noise addition.

3.2 Training Details

We reserve 80% of the data for training and 20% of the data for validation. Each model is trained for 10,000 iterations and after every 500 iterations, performance on the validation set is evaluated. The best-performing model on the validation set is used for generating the final prediction masks. Adam-optimizer is used along with a cyclic learning rate scheduler with a base learning rate of 0.0005 and a maximum learning rate of 0.003. A batch size of 1 is used and the base filters for OARFocalFuseNet, 3D-MSF, and SwinUNETR are 16, 16, and 48 respectively. We use an equally weighted combination of binary cross-entropy loss (see Eq. 6) and dice loss (see Eq. 7). Here, y is the ground truth value and $\hat{y}$ is the predicted value.

$$\mathcal{L}_{BCE} = (y - 1)\log(1 - \hat{y}) - y\log\hat{y}, \tag{6}$$

$$\mathcal{L}_{DSC} = 1 - \frac{2y\hat{y} + 1}{y + \hat{y} + 1}. \tag{7}$$

4   Results and Discussion

In this section, we present the comparisons of the selected baselines on our validation set. We report the quantitative evaluation in Table 1. Here, aggregated Dice Coefficient (DSC) is used as the metric for evaluating our results. From Table 1, we can observe that 3D-MSF obtains the highest aggregated DSC, highest classwise DSC on GTVp and GTVn. Meanwhile, OARFocalFuseNet is able to outperform SwinUNETR in terms of aggregated DSC and DSC obtained on GTVn. Thus, multi-scale fusion methodologies are able to report significant performance gains over other state-of-the-art (SOTA) methods and can be developed further for tumor segmentation in H&N CT/PET scans.

Table 1. Result comparison on HECKTOR 2022 Head and Neck Tumor segmentation challenge over our **validation set**. Aggregated DSC is reported along with class-wise DSC for GTVp and GTVn.

Method	DSC	GTVp	GTVn
SwinUNETR [5]	0.7828	0.7121	0.6364
OARFocalFuseNet [13]	0.7798	0.6898	0.6496
3D-MSF [13]	0.7951	0.7147	0.6706

5   Conclusion

In this paper, we compared the performance of three multi-scale fusion methodologies for H&N tumor segmentation to obtain accurate GTVn and GTVp. Our very recently proposed two multi-scale algorithms, originally designed for organ at-risk segmentation, were tuned for tumor segmentation from PET/CT scans obtained from the HECKTOR 2022 Challenge. We observed that under a supervised scenario, our proposed 3D-MSF and OARFocalFuseNet algorithms perform well on the HECKTOR 2022 H&N segmentation challenge. We plan to extend the multi-scale fusion strategy by introducing domain adaptation or generalization strategies within the framework to further advance the performance on HECKTOR 2022 H&N Segmentation Challenge.

Acknowledgment. This project is supported by the NIH funding: R01-CA246704 and R01-CA240639. The computations in this paper were performed on equipment provided by the Experimental Infrastructure for Exploration of Exascale Computing (eX3), which is financially supported by the Research Council of Norway under contract 270053.

References

1. Aerts, H.J., et al.: Decoding tumour phenotype by noninvasive imaging using a quantitative radiomics approach. Nat. Commun. **5**(1), 1–9 (2014)
2. Andrearczyk, V., et al.: Overview of the HECKTOR challenge at MICCAI 2021: automatic head and neck tumor segmentation and outcome prediction in pet/ct images. In: Andrearczyk, V., Oreiller, V., Hatt, M., Depeursinge, A. (eds.) HECKTOR 2021. LNCS, vol. 13209, pp. 1–37. Springer, Cham (2021). https://doi.org/10.1007/978-3-030-98253-9_1
3. Dosovitskiy, A., et al.: An image is worth 16×16 words: transformers for image recognition at scale. arXiv preprint arXiv:2010.11929 (2020)
4. Gu, J., et al.: HRViT: multi-scale high-resolution vision transformer. arXiv preprint arXiv:2111.01236 (2021)
5. Hatamizadeh, A., Nath, V., Tang, Y., Yang, D., Roth, H.R., Xu, D.: Swin UNETR: swin transformers for semantic segmentation of brain tumors in MRI images. In: Crimi, A., Bakas, S. (eds.) BrainLes 2021. LNCS, vol. 12962, pp. 272–284. Springer, Cham (2022). https://doi.org/10.1007/978-3-031-08999-2_22
6. Khan, S., Naseer, M., Hayat, M., Zamir, S.W., Khan, F.S., Shah, M.: Transformers in vision: a survey. ACM Comput. Surv. (CSUR) **54**(10s), 1–41 (2021)
7. Liu, Z., et al.: Swin transformer: hierarchical vision transformer using shifted windows. In: Proceedings of the IEEE/CVF International Conference on Computer Vision, pp. 10012–10022 (2021)
8. Lou, B., et al.: An image-based deep learning framework for individualising radiotherapy dose: a retrospective analysis of outcome prediction. Lancet Digit. Health **1**(3), e136–e147 (2019)
9. Oreiller, V., et al.: Head and neck tumor segmentation in PET/CT: the HECKTOR challenge. Med. Image Anal. **77**, 102336 (2022)
10. Ronneberger, O., Fischer, P., Brox, T.: U-Net: convolutional networks for biomedical image segmentation. In: Navab, N., Hornegger, J., Wells, W.M., Frangi, A.F. (eds.) MICCAI 2015. LNCS, vol. 9351, pp. 234–241. Springer, Cham (2015). https://doi.org/10.1007/978-3-319-24574-4_28
11. Srivastava, A., Chanda, S., Jha, D., Pal, U., Ali, S.: GMSRF-Net: an improved generalizability with global multi-scale residual fusion network for polyp segmentation. arXiv preprint arXiv:2111.10614 (2021)
12. Srivastava, A., et al.: MSRF-Net: a multi-scale residual fusion network for biomedical image segmentation. IEEE J. Biomed. Health Inform. **26**(5), 2252–2263 (2021)
13. Srivastava, A., Jha, D., Keles, E., Aydogan, B., Abazeed, M., Bagci, U.: An efficient multi-scale fusion network for 3D organ at risk (OAR) segmentation. arXiv preprint arXiv:2208.07417 (2022)
14. Wang, J., et al.: Deep high-resolution representation learning for visual recognition. IEEE Trans. Pattern Anal. Mach. **43**(10), 3349–3364 (2020)

Swin UNETR for Tumor and Lymph Node Segmentation Using 3D PET/CT Imaging: A Transfer Learning Approach

Hung Chu$^{(\boxtimes)}$, Luis Ricardo De la O Arévalo$^{(\boxtimes)}$, Wei Tang$^{(\boxtimes)}$,
Baoqiang Ma$^{(\boxtimes)}$, Yan Li$^{(\boxtimes)}$, Alessia De Biase$^{(\boxtimes)}$, Stefan Both$^{(\boxtimes)}$,
Johannes Albertus Langendijk$^{(\boxtimes)}$, Peter van Ooijen$^{(\boxtimes)}$,
Nanna Maria Sijtsema$^{(\boxtimes)}$, and Lisanne V. van Dijk$^{(\boxtimes)}$

University Medical Center Groningen (UMCG), 9700 RB Groningen, The Netherlands
{d.h.chu,l.r.de.la.o.arevalo,w.tang,b.ma,y.li05,a.de.biase,s.both,
j.a.langendijk,p.m.a.van.ooijen,n.m.sijtsema,l.v.van.dijk}@umcg.nl

Abstract. Delineation of Gross Tumor Volume (GTV) is essential for the treatment of cancer with radiotherapy. GTV contouring is a time-consuming specialized manual task performed by radiation oncologists. Deep Learning (DL) algorithms have shown potential in creating automatic segmentations, reducing delineation time and inter-observer variation. The aim of this work was to create automatic segmentations of primary tumors (GTVp) and pathological lymph nodes (GTVn) in oropharyngeal cancer patients using DL. The organizers of the HECKTOR 2022 challenge provided 3D Computed Tomography (CT) and Positron Emission Tomography (PET) scans with ground-truth GTV segmentations acquired from nine different centers. Bounding box cropping was applied to obtain an anatomic based region of interest. We used the Swin UNETR model in combination with transfer learning. The Swin UNETR encoder weights were initialized by pre-trained weights of a self-supervised Swin UNETR model. An average Dice score of 0.656 was achieved on a test set of 359 patients from the HECKTOR 2022 challenge. Code is available at: https://github.com/HC94/swin_unetr_hecktor_2022.

Aicrowd Group Name: RT_UMCG

Keywords: Head and neck cancer · Deep learning · Swin UNETR · HECKTOR 2022 · Radiotherapy · Tumor segmentation · Lymph node segmentation · Auto contouring · Image processing

1 Introduction

Head and neck cancers (HNC) are among the most common worldwide (5th leading cancer by incidence) [8]. Radiation therapy (RT) is pivotal in the treatment of HNC patients, however more than one out of four of all HNC patients in Europe did not receive RT due to limited trained personnel and equipment [9]. Accurate delineation of the tumor contour is important for delivering high dose in the tumor area without damaging surrounding normal tissues. However, delineation of the tumor contour is usually performed by experts and is susceptible to inter-observer variability. Treatment planning would benefit from automatic analysis of medical imaging data, as automatic segmenting tumors can reduce delineation time and interobserver variability.

In recent years, Deep Learning (DL) models have shown to be great potential for the medical field. More specifically, DL-based algorithms using fluorodeoxyglucose (FDG) Positron Emission Tomography (PET) and Computed Tomography (CT) as inputs have been explored in previous HECKTOR challenges for auto-segmenting GTV contour of the primary tumor, herewith showing promising results in terms of Dice scores [1,2].

The aim of this paper is to segment Head and Neck (H&N) primary tumors and lymph nodes in FDG-PET/CT images using a DL algorithm. We propose the Swin UNETR model, which showed top performance results for 3D semantic segmentation of brain tumors in Magnetic Resonance Imaging (MRI) images [4]. Furthermore, we performed transfer learning by using weights from a self-supervised Swin UNETR model [10].

2 Methods and Materials

2.1 Data

The training dataset available consisted of 524 HNC patients with histologically proven oropharyngeal cancer who underwent radiotherapy and/or radiochemotherapy treatment planning. The data was collected from seven different medical centers and provided by the organizers of the HECKTOR (HEad and neCK TumOR) 2022 challenge [1,7] (Table 1). For each patient a 3D FDG-PET scan, a 3D CT scan and the GTVp and GTVn segmentations (RTSTRUCT) were available. The GTVp and GTVn contours, used as ground-truth during training, were manually delineated by an annotator and cross checked by another annotator. Delineation guidelines were elaborated to ensure unification. The FDG-PET and low-dose non-contrast-enhanced CT images were acquired with combined PET/CT scanners. The independent test set (i.e. not used in model training) was a cohort of 359 HNC patients with FDG-PET and CT scans collected from three different centers (Table 1).

Table 1. Number of patients from each center.

	CHUM	CHUP	CHUS	CHUV	MDA	HGJ	HMR	USZ	CHB	*Total*
Training	56	72	72	53	198	55	18	0	0	524
Test	0	0	0	0	200	0	0	101	58	359

All files were provided in Nifti format. More information about medical data centers, scanners and data availability can be found at the following link: https://hecktor.grand-challenge.org/Data/.

2.2   Data Preprocessing

As data was collected by different centers, we preprocessed the data to obtain unification and adapted it to the input type required by our model. Firstly, we resampled the FDG-PET, CT and segmentations to an isotropic voxel spacing of $1 \times 1 \times 1\,\text{mm}^3$. The FDG-PET and CT scans were resampled with spline interpolation of degree 3, and the segmentations with nearest neighbor interpolation. Then we cropped a bounding box region using the automatic bounding box extraction algorithm from last year's HECKTOR challenge [3]: firstly, the brain is detected as the largest component containing SUV larger than 3, and secondly a rigid sized bounding box was placed at anatomic midpoints voxels in the x and y-axis, and at the lowest brain voxel in the z-axis. To increase the field of view, we increased the bounding box size from $144 \times 144 \times 144$ ($height\ (H) \times width\ (W) \times depth\ (D)$) to $192 \times 192 \times 192$. The FDG-PET and CT intensity values were expressed in Standard Uptake Value (SUV) and Hounsfield Units (HU) respectively, and were clipped between $[0, 25]$ SUV and $[-200, 400]$ HU. Lastly, we normalized the values to $[0, 1]$ as per $x_{norm} = \frac{x - x_{min}}{x_{max} - x_{min}}$, where $x_{min} = 0$, $x_{max} = 25$ SUV for the PET modality and $x_{min} = -200$, $x_{max} = 400$ HU for the CT modality.

2.3   Model Architecture

We used the Swin UNETR model [10]. An overview of the original Swin UNETR model architecture is depicted in Fig. 1. Firstly, the model projected the multi-modal input data into a 1D embedding sequence. Secondly, the embedding sequence was used as input for the Swin UNETR encoder, which was composed of a stack of Swin Transformer blocks. The output of each block was used in a U-Net style.

   We tailored the Swin UNETR model for our task. Our model accepted FDG-PET and CT as inputs with combined size $96 \times 96 \times 96 \times 2$, and generated a segmentation map of size $96 \times 96 \times 96 \times 3$ for background, GTVp and GTVn combined. The Swin UNETR encoder weights were initialized with pre-trained weights from a self-supervised Swin UNETR encoder that was trained on a cohort of 5050 CT scans from publicly available datasets [10]. The encoder was pre-trained on three different tasks: inpainting, contrastive learning and rotation prediction. To be able to use the pre-trained encoder weights, we used the same embedding size as the pre-trained model (i.e. 48 features). The reason for this is that an embedding of size 48 was used as input by the pre-trained encoder.

   An ensemble model was created from the seven models of a 7-fold cross-validation (CV) ('leave-one-center-out'-approach): one model from each CV fold. More specifically, for each patient in the test set we averaged the probability segmentation map of the seven models, and then discretized by applying arguments of the maxima (arg max) to obtain the ensemble segmentation map.

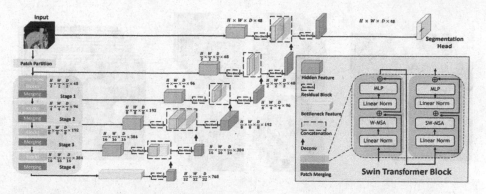

Fig. 1. Overview of the Swin UNETR model architecture, taken with permission from [10]. Firstly, the input data was projected into a 1D embedding sequence, which consequently was used as input for the encoder. A linear embedding of 48 features was used, same as the pre-trained model, to be able to use the pre-trained encoder weights.

2.4   Experiments

In each training iteration we randomly selected two fixed sized regions of size $96 \times 96 \times 96$ from a full input of size $192 \times 192 \times 192$. Since the majority of background voxels imbalanced the data, we selected the regions such that half of the all selected regions had a foreground (either from GTVp or GTVn) voxel in the center of the region, and the other half had a background voxel in the center of the region. The cropped regions were used as input batch of size two for the model. For model inferences we performed a sliding window approach as depicted in Fig. 2.

The model was trained for 200 epochs using the Dice + Cross-Entropy (DiceCE) loss function and the AdamW optimizer [5], and validated on the held-out fold with multi-class mean Dice score. The model weights were saved at the epoch with the highest validation score. The learning rate was updated using cosine annealing schedule with warm restarts [6]. Data augmentation techniques were adopted such as random translation, zooming, flipping, rotating and intensity shifting[1]. Each data augmentation technique was independently applied with 0.5 probability. The data augmentation and modeling were implemented using Project MONAI 0.9 in PyTorch 1.10. A comprehensive list of all training methodology is summarized in Table 2. The experiments were conducted on NVIDIA V100 GPU with 32 GB GPU memory.

2.5   Quantitative Evaluation

The experimental results were evaluated in terms of the aggregated Dice similarity coefficient $DSC_{agg} = \frac{2 \sum_i^N \sum_k \hat{y}_{i,k} \cdot y_{i,k}}{\sum_i^N \sum_k (\hat{y}_{i,k} + y_{i,k})}$, where N is the total number of test images, $y_{i,k}$ is the ground truth (either GTVp or GTVn) for voxel k of image i, and $\hat{y}_{i,k}$ is the prediction.

---

[1] https://docs.monai.io/en/stable/transforms.html.

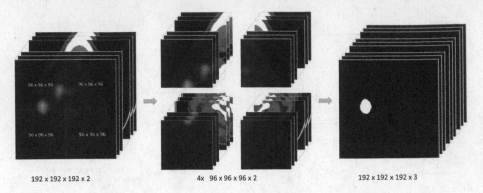

Fig. 2. Sliding window approach. Firstly, divide the full input of size 192 × 192 × 192 × 2 uniformly into four windows of size 96 × 96 × 96 × 2. Secondly, perform model inference on each windows. Finally, aggregate the output into a single segmentation map of size 192 × 192 × 192 × 3 for background, GTVp and GTVn combined.

Table 2. Training methodology and hyper-parameters.

Component	Value
Epochs	200
Batch size	2
Initial learning rate	$1e^{-4}$
Loss function	DiceCE
Optimizer	AdamW ($\beta_1 = 0.9$, $\beta_2 = 0.999$)
Scheduler	Cosine ($T_0 = 40$)
Weight decay	$1e^{-5}$
Data augmentation	Translating $[-10, 10]$, zooming $[90, 110]\%$, flipping, rotating $[-180°, 180°]$, intensity shifting $[-0.1, 0.1]$

3   Results

Table 3 presents the results of the 7-fold CV for each fold separately as well as the average over all folds. We observe that the DSC_{agg} of GTVp is always higher than that of GTVn. Moreover, the scores can differ significantly across folds.

For this challenge we submitted predictions from the ensemble model and the model from fold 5, which had the highest average DSC_{agg} in CV. Table 4 presents the results on the test set. The average Dice of GTVp and GTVn contouring is higher on the test set than the average CV result. Interestingly, we observed a significantly lower Dice score of GTVp segmentation on the test set than in CV, and the opposite for GTVn segmentation. Furthermore, the performance of the ensemble and fold 5 model are similar.

Table 3. Evaluation performance of each CV fold during training as well as the average over all folds.

	$DSC_{agg}\ GTVp$	$DSC_{agg}\ GTVn$	$\frac{DSC_{agg}\ GTVp\ +\ DSC_{agg}\ GTVn}{2}$
Fold 1	0.675	0.600	0.638
Fold 2	0.753	0.504	0.629
Fold 3	0.613	0.580	0.596
Fold 4	0.711	0.536	0.623
Fold 5	0.752	0.583	0.667
Fold 6	0.688	0.587	0.637
Fold 7	0.758	0.419	0.589
Average	0.707	0.582	0.626

Table 4. Evaluation performance of the ensemble model and the model from fold 5 on the test set.

	$DSC_{agg}\ GTVp$	$DSC_{agg}\ GTVn$	$\frac{DSC_{agg}\ GTVp\ +\ DSC_{agg}\ GTVn}{2}$
Ensemble	0.642	0.670	0.656
Fold 5	0.633	0.673	0.653

4   Discussion and Conclusion

In this paper we proposed to use Swin UNETR model in conjunction with transfer learning. We combined FDG-PET and CT images into a single input for our end-to-end model.

The self-supervised pre-trained Swin UNETR model was trained on CT scans and no FDG-PET imaging. Therefore the pre-trained weights may not be helpful for the FDG-PET modality. Another limitation is the computational time: training the model for 200 epochs for a single CV fold iteration took about two days. Therefore we did not do any hyperparameter tuning. With one GPU we recommend to apply training-validation split instead of CV and train for more than 200 epochs, because the validation performance in the CV interations was still improving at 200 epochs.

Also, we observed unexpected values in the provided training and test data. The CT input intensity values (i.e. after cropping the bounding boxes and resampling) in the training and test data was in the interval $[-17.200, 32.636]$ and $[-10.223, 38.010]$, respectively. However, CT intensity values in HU should be in the interval $[-1.024, 3.000]$. In fact, only 6% of the training patients complied with that, and 66% of the training patients had CT intensity value in $[-4.000, 4.000]$. This holds similarly for the test data. These findings suggest that most CT scans were not represented in HU. Different data normalization techniques should have been applied to obtain data unification. We did not deal with these issues due to late discovery of these issues and therefore lack of time.

The ensemble and fold 5 model have similar Dice scores on the test set. On top of that, DSC_{agg} of GTVp is higher than that of GTVn in CV, while this is opposite for the test set. Therefore we suspect that the data across centers may differ significantly, possibly due to different PET/CT scanners, and require additional data preprocessing to obtain data unification.

For future work we suggest to use pre-trained weights trained on FDG-PET imaging, perform hyperparameter tuning, train for more than 200 epochs, and improve uniformity of the data modalities.

Acknowledgements. We thank the Center for Information Technology of the University of Groningen for their support and for providing access to the Peregrine high performance computing cluster.

References

1. Andrearczyk, V., et al.: Overview of the HECKTOR challenge at MICCAI 2022: automatic head and neck tumor segmentation and outcome prediction in PET/CT. In: Andrearczyk, V., Oreiller, V., Hatt, M., Depeursinge, A. (eds.) HECKTOR 2022. LNCS, vol. 13626, pp. 1–30. Springer, Cham (2023)
2. De Biase, A., et al.: Skip-SCSE multi-scale attention and co-learning method for oropharyngeal tumor segmentation on multi-modal PET-CT images. In: Andrearczyk, V., et al. (eds.) HECKTOR 2021. LNCS, vol. 13209, pp. 109–120. Springer, Cham (2022). https://doi.org/10.1007/978-3-030-98253-9_10
3. Andrearczyk, V., Oreiller, V., Depeursinge, A.: Oropharynx detection in PET-CT for tumor segmentation. In: Irish Machine Vision and Image Processing (2020)
4. Hatamizadeh, A., et al.: Swin UNETR: swin transformers for semantic segmentation of brain tumors in MRI images (2022). https://doi.org/10.48550/arXiv.2201.01266
5. Loshchilov, I., et al.: Decoupled weight decay regularization (2017). https://doi.org/10.48550/arXiv.1711.05101
6. Loshchilov, I., et al.: SGDR: stochastic gradient descent with warm restarts (2016). https://doi.org/10.48550/arXiv.1608.03983
7. Oreiller, V., et al.: Head and neck tumor segmentation in PET/CT: the HECKTOR challenge. Med. Image Anal. **77**, 102336 (2022)
8. Parkin, D.M., et al.: Global cancer statistics, 2002. CA Cancer J. Clin. **55**(2), 74–108 (2005)
9. Lievens, Y.: Provision and use of radiotherapy in Europe. Mol. Oncol. **14**(7), 1461–1469 (2020). https://doi.org/10.1002/1878-0261.12690
10. Tang, Y., et al.: Self-supervised pre-training of swin transformers for 3D medical image analysis (2021). https://doi.org/10.48550/arXiv.2111.14791

Simplicity Is All You Need: Out-of-the-Box nnUNet Followed by Binary-Weighted Radiomic Model for Segmentation and Outcome Prediction in Head and Neck PET/CT

Louis Rebaud[1,2]([✉]), Thibault Escobar[1,3]([✉]), Fahad Khalid[1], Kibrom Girum[1], and Irène Buvat[1]

[1] Laboratory of Translational Imaging in Oncology, U1288, Institut Curie, Inserm, Université Paris-Saclay, Orsay, France
louis.rebaud@gmail.com, thibescobar@gmail.com
[2] Siemens Healthcare SAS, Saint Denis, France
[3] DOSIsoft SA, Cachan, France

Abstract. Automated lesion detection and segmentation might assist radiation therapy planning and contribute to the identification of prognostic image-based biomarkers towards personalized medicine. In this paper, we propose a pipeline to segment the primary and metastatic lymph nodes from fluorodeoxyglucose (FDG) positron emission tomography and computed tomography (PET/CT) head and neck (H&N) images and then predict recurrence free survival (RFS) based on the segmentation results. For segmentation, an out-of-the-box nnUNet-based deep learning method was trained and labelled the two lesion types as primary gross tumor volume (GTVp) and metastatic nodes (GTVn). For RFS prediction, 2421 radiomic features were extracted from the merged GTVp and GTVn using the pyradiomics package. The ability of each feature to predict RFS was measured using the C-index. Only the features with a C-index greater than C_{min}, hyperparameter of the model, were selected and assigned a +1 or −1 weight as a function of how they varied with the recurrence time. The final RFS probability was calculated as the mean across all selected feature z-scores weighted by their +/−1 weight. The fully automated pipeline was applied to the data provided through the HECKTOR 2022 MICCAI challenge. On the test data, the fully automated segmentation model achieved 0.777 and 0.763 Dice scores on the primary tumor and lymph nodes respectively (0.770 on average). The binary-weighted radiomic model yielded a 0.682 C-index. These results allowed us to rank first for outcome prediction and fourth for segmentation in the challenge. We conclude that the proposed fully-automated pipeline from segmentation to outcome prediction using a binary-weighted radiomic model competes well with more complicated models. Team: LITO.

Keywords: Medical imaging · Survival prediction · Segmentation · FDG PET/CT · Head and neck · Machine learning

© The Author(s), under exclusive license to Springer Nature Switzerland AG 2023
V. Andrearczyk et al. (Eds.): HECKTOR 2022, LNCS 13626, pp. 121–134, 2023.
https://doi.org/10.1007/978-3-031-27420-6_13

1   Introduction

Quantitative medical image analysis assists in patient staging, treatment planning and monitoring, and overall patient management. In head and neck (H&N) cancer, fluorodeoxyglucose (FDG) positron emission tomography combined with computed tomography (PET/CT) is a modality of choice for initial staging and patient follow-up and contributes to radiation therapy planning. Indeed, H&N cancer primary treatment mostly relies on radiotherapy and requires target volume delineation of the gross primary tumor volume (GTVp) and cancer node volumes (GTVn) on PET/CT images, which is time-consuming and prone to intra/inter-observer variabilities. Automated segmentation might allow radiation oncologists to optimize the treatment plan in a shorter time while improving reproducibility. In addition, the prediction of the risk of relapse based on medical images could help identify patients for whom treatment intensification and close monitoring might be needed.

In the recent years, machine learning (ML) and radiomics have been instrumental in advancing automated image segmentation and building predictive models. Yet, the diversity of datasets on which methods are designed and tested makes it difficult to compare their performance and determine which one is best suited in a particular context. Given the possible sensitivity of automated segmentation and predictive models to image quality, multi-center evaluation of these methods is absolutely needed before considering clinical deployment.

Challenges offer unique opportunities for testing and comparing the performance of different methods on a common database using large multi-center datasets. The HEad and neCK TumOR (HECKTOR) challenges organized as part of MICCAI aims at establishing best-performing methods for segmentation and prediction tasks [1,2]. In 2022, the HECKTOR challenge first task was to automatically segment the H&N GTVp and GTVn from FDG PET/CT images. The second task consisted in automatically predicting patient outcomes from a PET/CT image, with or without clinical information, with PET/CT images and clinical information collected from nine different centers.

Several contributions to the automated segmentation in the context of H&N cancer have already been published. Guo et al. proposed a modified U-net approach using dense blocks and reached 0.71 average Dice score on a public multi-center dataset of 250 PET/CT H&N patients [3]. Their study also showed that combining PET and CT in two channels substantially increased the segmentation performance compared to using PET (0.64 average Dice score) or CT (0.31 average Dice score) alone. Ren et al. compared several modality combinations including PET, CT, and magnetic resonance imaging (MRI) on a multi-center dataset of 153 patients for deep learning tumor segmentation using a U-net approach [4]. All combinations including PET provided similar results (0.72 to 0.74 Dice score), while the anatomic-only combination (CT and MRI) led to a lower score (0.58). More generally, automated medical image segmentation is currently dominated by deep convolutional neural networks (CNN) [5–7]. Most methods rely on U-net based approaches with several context-specific changes in model architecture, training scheme, and data pre- or post-processing. In HECKTOR

2021 challenge, the best-performing segmentation method used a tuned nnUNet with squeeze and excitation (SE) layers on fused PET and CT images, yielding a 0.779 Dice score on primary tumor [7,8].

Similarly, models have been proposed to predict patient outcome from PET/CT images in H&N cancer (e.g., [9,10]). In HECKTOR 2021, two different methods performed best at predicting the progression free survival [11,12]. Both were based on a CNN trained on unsegmented images using large bounding boxes, and achieved 0.720 and 0.694 C-index on the test data respectively. A logistic model based on radiomic features calculated from the segmented tumor region also performed well with a 0.683 C-index [13].

This paper presents our simple and efficient pipeline for fully automatic segmentation and outcome prediction method and its performance on the HECKTOR 2022 challenge data. For the segmentation task, we adapted the publicly available nnUNet deep learning framework to detect and segment the H&N primary tumor (GTVp) and nodal gross tumor volumes (GTVn) [7]. For the prediction task, we introduce a novel binary-weighted model operating on radiomic features calculated from the tumor regions automatically segmented in the previous step. The evaluation was conducted on the HECKTOR 2022 challenge data and the models are publicly available.

2   Materials and Methods

Here, we describe our proposed fully-automatic end-to-end framework to segment lesions and predict outcome from 18F-FDG PET/CT images (Fig. 1). First, a well established out-of-the-box nnUNet deep learning method was trained to segment and label the GTVp and GTVn [7]. From the segmented GTVp and GTVn regions, we extracted radiomic features. We then applied the binary-weighted model to rank the patients as a function of their recurrence free survival.

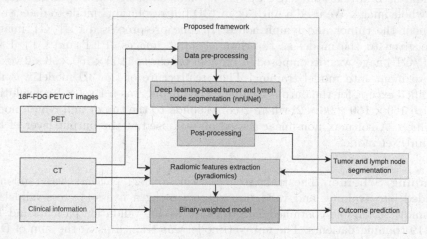

Fig. 1. Proposed framework: schematic representation of the fully-automatic pipeline from segmentation to outcome prediction.

2.1 Data

To develop and evaluate the proposed method, we used the HECKTOR 2022 data that included FDG PET/CT images, clinical and survival data of 524 patients from 7 centers for training and PET/CT and clinical data only of 359 patients from 3 centers for blind testing of the models [1,2]. In the training data, reference segmentations of the primary tumor (GTVp) and metastatic nodes (GTVn) were provided. Train and test PET/CT scans were provided with 9 clinical features with some missing values: gender, age, weight (1.23% missing values), tobacco ($0 = no$, $1 = yes$) (61.1% missing), alcohol ($0 = no$, $1 = yes$) (68.5% missing), performance status (56.0% missing), human papillomavirus (HPV) status ($0 = no$, $1 = yes$) (35.2% missing), surgery ($0 = no$, $1 = yes$) (38.7% missing), and chemotherapy ($0 = no$, $1 = yes$). RFS was provided for 488 patients in the train set, and 339 patients of the test set for whom the outcome was known were concerned by the outcome prediction (task 2).

Data Pre-processing: The training CT images had an original median voxel-size of $0.976 \times 0.976 \times 2.798$ mm^3 and the PET images had median voxel-size of $4.000 \times 4.000 \times 3.270$ mm^3. All PET/CT images and corresponding segmentations were resampled to $2.0 \times 2.0 \times 2.0$ mm^3. CT and PET images were resampled using a third-order spline. The segmentation mask was resampled using nearest neighbor interpolation.

2.2 Tumor and Lymph Node Segmentation

Deep Learning Model: All CT images were clipped between 0.5^{th} and 99.5^{th} percentile of the Hounsfield Units (HU) intensity values and normalized using z-score based on all training images. To favor contrast-based features in PET, PET standardized uptake values (SUV) were normalized using z-score patient-wise on the whole image. We used a nnUNet in "3D full resolution" mode to detect and segment the tumor and lymph nodes [7]. The pre-processed PET/CT images were given to the model as two-channel input images (PET and CT). Each PET/CT image was decomposed in random patches of $160 \times 160 \times 96 \times 2$ voxels before input into model training. The architecture of the 3D model was not modified except for the output channel. The output was a $1 \times 1 \times 1$ convolution of size $160 \times 160 \times 96 \times 2$, where 2 corresponds to the tumor and lymph nodes channels. A softmax non-linear activation was used at the output layer of the 3D nnUNet model.

Training Scheme: The train set consisting of 524 patients was randomly divided into training and validation subsets using a five-fold cross-validation technique. Each fold contained data from 104 or 105 validation patients and 420 or 419 training patients. The nnUNet model was trained using the sum of Dice and cross-entropy losses. The initial number of feature maps in the architecture was 32. Performance assessment and post-processing strategy were determined based on the five-fold cross-validation with 1000 epochs training, with an initial learning rate of 0.01 and a scheduler weight decay of $3e^{-5}$. We selected a batch size of two. Other hyper-parameter settings, including data augmentation techniques, were the default settings of nnUNet. Implementation was done in Pytorch and training was performed using four GPUs: three NVIDIA Quadro RTX 5000 with 16 GB and one NVIDIA RTX A6000 with 49 GB GPU memory. On average, the training time was 141 s per epoch on NVIDIA Quadro RTX 5000 and 82 s on NVIDIA RTX A6000.

Post-processing: The segmentation output of the deep learning model had a $2 \times 2 \times 2$ mm^3 voxel spacing. It was then resampled into the corresponding original CT spacing. Then, a median filter with a $3 \times 3 \times 3$ voxel kernel size was applied to smooth out the staircase effect.

Prediction on the Test Set: For predictions on the test set, three strategies were used. First we ensembled the five models trained during cross-validation. Second, a bagging strategy was adopted to increase the number of ensembled models to nine. Nine models were trained on random samples of size equal to the whole dataset drawn with replacement (i.e. bootstrap samples). The predictions from the models were then aggregated using majority voting. Nine was the maximum number of models we could train on our GPUs for this strategy within the allotted time of the challenge. Finally, we increased the number of epochs to 1500 and trained only one model on the whole dataset.

2.3   Outcome Prediction

Our prediction model was based on engineered radiomic features extracted from the tumor regions segmented using the automated approach described in Sect. 2.2. These features were then analyzed using an original approach yielding what we call a binary-weighted model.

Radiomic Features Extraction: We used the segmentation mask produced by the deep learning model described in Sect. 2.2. Primary tumor and lymph node regions were merged as a single "lesion" mask. To make the model less sensitive to potential segmentation errors, multiple masks were created from this binary lesion mask:

- Original lesion mask
- Smallest bounding box enclosing all the lesions
- Lesion mask refined by removing all voxels in which SUV was less than 2.5
- Lesion mask refined by removing all voxels in which SUV was less than 4
- Lesion mask re-segmented with a threshold of 40% of global SUVmax
- Lesion mask dilated by 1mm (resp 2, 4, 8 and 16 mm)
- A 2mm (resp 4, 8 mm) thick shell surrounding each connected component of the lesion mask

For each of these 13 masks, 93 radiomic features were computed on the PET image and 93 on the CT image with pyradiomics [14]. These features were the default features from pyradiomics, composed of features reflecting the ROI shape, and the signal intensity and texture. A fixed-bin size of 0.3 SUV units was used for PET images and 10 HU for the CT. Three handcrafted features were added: the number of tumor masses, the number of lymph nodes, and a binary variable indicating whether the scan was a whole-body scan or included only the H&N region. This was determined by calculating the length of the scan in the axial direction from the image volume. Used together with the provided nine clinical features, this pipeline produced 2430 features.

Binary-weighted Model: From the literature and our experience, we hypothesize that it is difficult to accurately estimate biomarker importance in outcome prediction. For instance, Adams et al. found the national comprehensive cancer network international prognostic index to be more predictive of progression free survival than whole-body total metabolic tumor volume in diffuse large B-cell lymphoma, while Cottereau et al. observed the opposite [15,16]. Indeed, noise in the data, censoring of the target, e.g. progression free survival, and relatively low number of training samples might increase the risk of biased estimation of the feature weights. To mitigate this effect, we propose to reduce the learned information to the bare minimum and only estimate a sign to be assigned to each feature for estimating the target. This is the core mechanism of the introduced binary-weighted model.

Definition: Our training dataset includes N samples and M features. Many radiomic features are highly correlated. To comply with the basic assumption of our binary-weighted model, only one among a set of correlated features should be kept because if they are all input to the model, this will artificially give a large weight to the information reflected by the feature. We thus perform feature selection by calculating the absolute value of the Pearson correlation coefficient for all pairs of features. A threshold ρ is used to set the value above which two features are deemed too correlated. In such case, one of the two features is randomly selected and dropped.

Let's C_{index} be the Harrell's concordance index [17]. Each feature x_i is evaluated on its ability to correctly predict the target value y with:

$$c_i = C_{index}(x_i, y) \tag{1}$$

To reduce the risk of wrong estimation of the sign, the features with $|c_i| < C_{min}$ are dropped, where $|c_i| = max\{1 - c_i, c_i\}$ and C_{min} is a hyperparameter in $[0.5, 1]$. The remaining features are assigned a sign as follows:

$$s_i = \begin{cases} +1, & \text{if } c_i \geq 0.5 \\ -1, & otherwise \end{cases} \tag{2}$$

A normalization step is necessary to scale the feature values to the same range. Otherwise, features with large absolute values would have a higher weight in the final prediction. To do so, the model computes the z-score of each feature:

$$z_i = \frac{x_i - \mu_i}{\sigma_i} \tag{3}$$

where μ_i and σ_i are the mean and standard deviation of x_i in the train set. The estimate $\hat{y}$ of the target y is computed with:

$$\hat{y} = \frac{1}{M} \sum_i^M s_i \times z_i \tag{4}$$

The computation of $\hat{y}$, μ_i and σ_i are done by ignoring the missing values of the dataset. This allows the model to use features with missing values.

Here, C_{min} and ρ are the only two hyperparameters of the model.

Curse of Dimensionality: The curse of dimensionality is a phenomenon where we observe a loss in performance of ML models when too many features are given as an input. This especially occurs in medical datasets when the data are high-dimensional and the number of samples is low [18]. We hypothesize that the binary-weighted model is resilient to this phenomenon. We tested this hypothesis on the train set of the HECKTOR dataset by gradually increasing the number of features input to the model.

Ensembling: To produce a more precise and stable estimate $\hat{y}$, a bagging strategy was adopted as described in Sect. 2.2. An ensemble of E binary-weighted models were trained, each model being trained on a random sample of size N of the training data drawn with replacement. Each model also randomly selected F features to work with. The models were trained on their bootstrap sample from the train set and predicted $\hat{y}$ on the test set. The E predictions from the E models were then aggregated with the median. F is a hyperparameter of the ensemble model. Our experiments on the train set suggested that the higher E, the better the performance. We used $E = 10^5$ on the test set, a number large enough to ensure good results while keeping computational cost reasonable.

Cross-validation: To evaluate a model from the train set, we used a two-hundred-fold Monte Carlo cross-validation with a validation set of size $0.5 \times N$ (CV). This large number of folds was used to ensure precise comparison of the numerous tested algorithms, with reproducible results. The model prediction on the validation set was evaluated with Harrell's C-index. The average score and its confidence interval were reported.

Hyperparameters Optimization: The ensemble model has 3 hyperparameters: F, C_{min} and ρ. To determine the best hyperparameter set, random search was used. 1000 hyperparameter sets were randomly drawn and evaluated using CV. The hyperparameter sets were then ranked by their CV scores. To reduce the risk of overfitting the hyperparameter choice on the train set, the B best hyperparameter sets were selected, and for the prediction on the test set, an ensemble model was trained with each binary-weighted model randomly selecting a hyperparameter set from the selected B. The B value was optimized with an additional CV. Three bagged models were evaluated in the train and test sets of the HECKTOR challenge. While similar, each model used more and more hyperparameter sets in its random search, each time increasing the probability of overfitting on the train set. The number of hyperparameter sets tested was increased gradually through the 3 attempts given to the participating teams.

Feature Importance: While the binary-weighted model only gives weights of -1 or $+1$, after bagging, an approximation of feature importance can be computed by taking the average sign of each feature across all models. Feature importance was determined on the train set of HECKTOR.

3 Results

3.1 Segmentation Evaluation

In this section, except for the visual evaluation where it was assessed patient-wise, the Dice score was always computed on pseudo-volumes of the validation sets during cross-validation (aggregated Dice score).

Cross-validation: The Dice score across all images through the cross-validation was 0.850 for GTVp and 0.789 for GTVn (0.821 on average). For thorough comparison, Table 1 reports the Dice score across the different centers of acquisition.

Table 1. Dice scores for primary tumor and lymph node segmentation across the different centers evaluated on a five-fold cross-validation on the train set.

Center	Nb of patients	GTVp Dice	GTVn Dice	Average Dice
CHUP	72	0.868	0.687	0.778
CHUV	53	0.823	0.781	0.803
MDA	198	0.821	0.813	0.817
HMR	18	0.846	0.811	0.829
CHUS	72	0.865	0.805	0.835
CHUM	56	0.849	0.831	0.840
HGJ	55	0.883	0.829	0.856
All	**524**	**0.850**	**0.789**	**0.821**

Test: Table 2 displays the class-specific Dice scores for our three submitted models for evaluation on the test set. The model trained on all training data for 1500 epochs achieved the highest scores (highlighted in bold).

Table 2. Dice scores from our 3 methods on the test set of HECKTOR.

Method	GTVp Dice	GTVn Dice	Average Dice
Ensembled 5 folds	0.778	0.761	0.769
Bagging 9 samples	0.779	0.759	0.769
Whole train set	**0.777**	**0.763**	**0.770**

3.2   Qualitative Assessment

PET/CT images, ground truth and predicted segmentations are shown in Fig. 2 for 5 patients. The examples were selected based on the Dice scores. The top two rows display high Dice scoring patients (average Dice 0.922 and 0.910 respectively), the third row a patient with an average score (0.761), while the fourth (0.303) and fifth (0.000) rows display patients with the lowest scores.

Results for patients (1) and (2) were very satisfactory. In patient (3), the model accurately identified the two nodes and the tumor but missed some voxels, especially at the sharp edges. In patient (4), false positive node voxels were labeled by the model (not shown in the figure because not in the slice). Last, patient (5) shows an example of accurate detection and segmentation but with complete class mismatch. The green contour representing the tumor is precisely delineated by the model but labelled as a node, as shown by the pink predicted contour, yielding a Dice equal to zero.

3.3   Performance of the Outcome Prediction Model

Table 3 shows the results of the different models tested during the challenge. A binary-weighted model without bagging was evaluated only on the train set and not submitted because its performances were below the bagged models on the train set. The performance of the three submitted bagged models is correlated with the number of hyperparameter sets evaluated on the train set. The best model was the one which had the most extensive search of hyperparameters.

3.4   Resilience to the Curse of Dimensionality

Figure 3 shows the result of the experiment using the train set to test our hypothesis stating that binary-weighted models do not suffer from the curse of dimensionality. The performance plateaued when increasing the number of features used by the model up to the maximum number of available features.

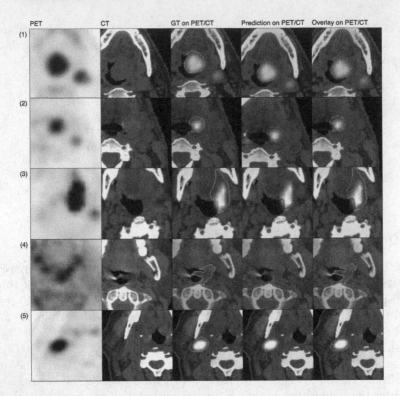

Fig. 2. Examples of PET/CT images, ground truth and predicted segmentation for five patients from the validation sets of the five-fold cross-validation. Green and blue ground truth contours correspond to tumor and lymph node respectively. Red and pink contours correspond to the predicted segmentation for tumor and lymph node. (Color figure online)

3.5   Feature Importance

The importance of the clinical and some representative radiomic features evaluated on the train set is presented in Fig. 4. The error bars are not shown because by construction of the model, they are unnecessary (the higher the absolute value, the lower the standard deviation).

4   Discussion

4.1   Segmentation

Our segmentation method was inspired by Xie and Peng [8] using Isensee et al. [7] framework. Our choice of not using the SE layers and keep PET and CT separated as two channels was based on the intuition that approaching the problem in a straightforward way would increase its robustness. Overall, our segmentation results were satisfactory, ranking fourth in the challenge with 0.770 average Dice, compared to the 0.788 Dice achieved by the winner.

Table 3. C-index and number of hyperparameters searched for the prediction models evaluated on the train and test set of the HECKTOR challenge. On the train set, the mean C-index over the CV is reported as well as the confidence interval (CI).

Model	CV C-index train set (CI)	C-index test set	Nb tested sets of hyperparameters
Binary-weighted	0.645 (0.585 − 0.707)		10
Binary-weighted bagged	0.668 (0.605 − 0.730)	0.670	10
Binary-weighted bagged	0.675 (0.613 − 0.731)	0.673	100
Binary-weighted bagged	**0.688 (0.642 − 0.732)**	**0.682**	1000

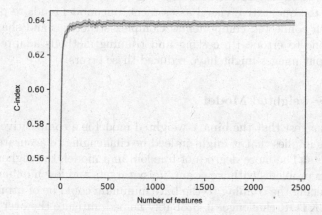

Fig. 3. Cross-validated C-index of a binary-weighted model (not bagged) when increasing the number of features. The features and hyperparameters were selected randomly.

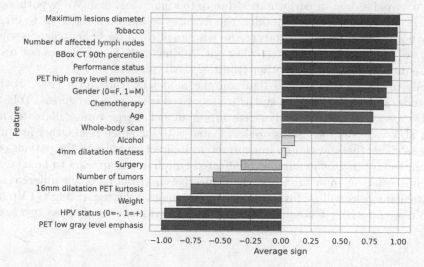

Fig. 4. Importance of the clinical and representative radiomic features. A positive value (red) shows a positive correlation with the risk and a negative value (blue) is a negative correlation. The higher the absolute value of the average sign, the more important the feature. "Whole-body scan" is 1 if the scan is whole-body or 0 if only H&N. (Color figure online)

Although the centers had different numbers of patients, Dice scores were consistently lower for lymph nodes than for primary tumor in all centers, demonstrating they are more difficult to segment. Mislabelling of node regions as seen in Fig. 2 decreased Dice value although contours were accurately delineated. One way to address this mislabelling could be to set higher weight to the lymph node class in the loss function.

According to our test results, the deployment strategy did not have a big impact on performance. Indeed, ensembling the cross-validation models, using a bagging strategy while increasing the number of models, or training only one model on the whole dataset, led to very similar performance.

Based on the qualitative visual assessment, our model tends to perform better on smooth connected components. Complex structures and sharp contours are more prone to errors. Processing and training methods adapted to higher resolution input images might have reduced these errors.

4.2   Binary-weighted Model

Our results suggest that the binary-weighted model is a competitive and robust method. This implies that it might indeed be challenging to accurately estimate feature weights. The more degrees of freedom in a model, the higher the risk of overfitting. In problems with weak and noisy targets and low number of training samples, reducing the training to the bare minimum could be of utmost interest. For the HECKTOR challenge, it probably helped mitigate the overfitting.

Figure 3 shows that the binary-weighted model does not suffer from the curse of dimensionality. The vast majority of ML algorithms need some feature selection to avoid a drop in performance due to too many features. We hypothesized that in our binary-weighted model, the features would work together to cancel their noise and biases, analogous to the wisdom of the crowd phenomenon where errors of individuals cancel each other out. Adding more features does not result in loss in performance as in other traditional ML methods.

Features importance shed light on the model interpretation (Fig. 4). For instance, a high performance status is associated with worse prognosis. Tobacco is also associated with a higher risk in our model. Large tumor diameter and high SUV values in the lesions are associated with increased risk. Other features, such as chemotherapy, can be interpreted as indirect measure of the patient condition. Interestingly, the number of affected lymph nodes appears to be a strong prognostic factor. In future work, the respective contribution of the different segmentation masks will be investigated. More importantly, separating GTVp and GTVn would make it possible to assess the individual role of these two lesion types.

5   Conclusions

We proposed a new, fully automated framework to predict outcomes in H&N patients from a given PET/CT image and clinical information. It involves deep learning-based GTVp and GTVn segmentation, radiomic feature extraction, and outcome prediction. Our pipeline including the novel binary-weighted radiomic model outperformed other methods for outcome prediction while providing accurate segmentation, ranking first for prediction and fourth for segmentation in the HECKTOR 2022 challenge. The number of lymph nodes was one of the prognostic features, highlighting the importance of lymph node segmentation for predicting the outcome in H&N cancer.

We created an easy-to-use package for the binary-weighted model, called Individual Coefficient Approximation for Risk Estimation (ICARE). The code is publicly available at: github.com/Lrebaud/ICARE.

References

1. Oreiller, V., et al.: Head and neck tumor segmentation in PET/CT: the HECKTOR challenge. Med. Image Anal. **77**, 102336 (2022)
2. Andrearczyk, V., et al.: Overview of the HECKTOR Challenge at MICCAI 2022: automatic head and neck tumor segmentation and outcome prediction in PET/CT. In: Head and Neck Tumor Segmentation and Outcome Prediction (2023)
3. Guo, Z., et al.: Gross tumor volume segmentation for head and neck cancer radiotherapy using deep dense multi-modality network. Phys. Med. Biol. **64**(20), 205015 (2019)
4. Ren, J., et al.: Comparing different CT, PET and MRI multi-modality image combinations for deep learning-based head and neck tumor segmentation. Acta Oncol. **60**(11), 1399–1406 (2021)
5. Menze, B.H., et al.: The multimodal brain tumor image segmentation benchmark (BRATS). IEEE Trans. Med. Imaging **34**(10), 1993–2024 (2014)
6. Antonelli, M., et al.: The medical segmentation decathlon. Nat. Commun. **13**(1), 1–13 (2022)
7. Isensee, F., et al.: nnU-Net: a self-configuring method for deep learning-based biomedical image segmentation. Nat. Methods **18**(2), 203–211 (2021)
8. Xie, J., Peng, Y.: The head and neck tumor segmentation based on 3D U-Net. In: Andrearczyk, V., Oreiller, V., Hatt, M., Depeursinge, A. (eds.) Head and Neck Tumor Segmentation and Outcome Prediction. HECKTOR 2021. LNCS, vol. 13209. Springer, Cham (2022). https://doi.org/10.1007/978-3-030-98253-9_8
9. Vallières, M., et al.: Radiomics strategies for risk assessment of tumour failure in head-and-neck cancer. Sci. Rep. **7**(1), 1–14 (2017)
10. Diamant, A., et al.: Deep learning in head & neck cancer outcome prediction. Sci. Rep. **9**(1), 1–10 (2019)
11. Saeed, N., Al Majzoub, R., Sobirov, I., Yaqub, M.: An Ensemble Approach for Patient Prognosis of Head and Neck Tumor Using Multimodal Data. In: Andrearczyk, V., Oreiller, V., Hatt, M., Depeursinge, A. (eds.) Head and Neck Tumor Segmentation and Outcome Prediction. HECKTOR 2021. LNCS, vol. 13209. Springer, Cham (2022). https://doi.org/10.1007/978-3-030-98253-9_26
12. Naser, M.A., et al.: Progression free survival prediction for head and neck cancer using deep learning based on clinical and PET/CT imaging data. In: Andrearczyk, V., Oreiller, V., Hatt, M., Depeursinge, A. (eds.) Head and Neck Tumor Segmentation and Outcome Prediction. HECKTOR 2021. LNCS, vol. 13209. Springer, Cham (2022). https://doi.org/10.1007/978-3-030-98253-9_27

13. Salmanpour, M.R., et al.: Advanced automatic segmentation of tumors and survival prediction in head and neck cancer. In: Andrearczyk, V., Oreiller, V., Hatt, M., Depeursinge, A. (eds.) Head and Neck Tumor Segmentation and Outcome Prediction. HECKTOR 2021. LNCS, vol. 13209. Springer, Cham (2022). https://doi.org/10.1007/978-3-030-98253-9_19

14. Griethuysen, V., et al.: Computational radiomics system to decode the radiographic phenotype. Can. Res. **77**(21), e104–e107 (2017)

15. Adams, H.J.A., et al.: Prognostic superiority of the national comprehensive cancer network international prognostic index over pretreatment whole-body volumetric-metabolic FDG-PET/CT metrics in diffuse large B-cell lymphoma. Eur. J. Haematol. **94**(6), 532–539 (2015)

16. Cottereau, A.-S., et al.: Risk stratification in diffuse large B-cell lymphoma using lesion dissemination and metabolic tumor burden calculated from baseline PET/CT. Ann. Oncol. **32**(3), 404–411 (2021)

17. Harrell, F.E., Jr., et al.: Multivariable prognostic models: issues in developing models, evaluating assumptions and adequacy, and measuring and reducing errors. Stat. Med. **15**(4), 361–387 (1996)

18. Berisha, V., et al.: Digital medicine and the curse of dimensionality. NPJ Digit. Med. **4**(1), 1–8 (2021)

Radiomics-Enhanced Deep Multi-task Learning for Outcome Prediction in Head and Neck Cancer

Mingyuan Meng[1] , Lei Bi[1]([⊠]) , Dagan Feng[1,2] , and Jinman Kim[1]

[1] School of Computer Science, The University of Sydney, Sydney, Australia
lei.bi@sydney.edu.au
[2] Med-X Research Institute, Shanghai Jiao Tong University, Shanghai, China

Abstract. Outcome prediction is crucial for head and neck cancer patients as it can provide prognostic information for early treatment planning. Radiomics methods have been widely used for outcome prediction from medical images. However, these methods are limited by their reliance on intractable manual segmentation of tumor regions. Recently, deep learning methods have been proposed to perform end-to-end outcome prediction so as to remove the reliance on manual segmentation. Unfortunately, without segmentation masks, these methods will take the whole image as input, such that makes them difficult to focus on tumor regions and potentially unable to fully leverage the prognostic information within the tumor regions. In this study, we propose a radiomics-enhanced deep multi-task framework for outcome prediction from PET/CT images, in the context of HEad and neCK TumOR segmentation and outcome prediction challenge (HECKTOR 2022). In our framework, our novelty is to incorporate radiomics as an enhancement to our recently proposed Deep Multi-task Survival model (DeepMTS). The DeepMTS jointly learns to predict the survival risk scores of patients and the segmentation masks of tumor regions. Radiomics features are extracted from the predicted tumor regions and combined with the predicted survival risk scores for final outcome prediction, through which the prognostic information in tumor regions can be further leveraged. Our method achieved a C-index of 0.681 on the testing set, placing the 2nd on the leaderboard with only 0.00068 lower in C-index than the 1st place.

Keywords: Outcome prediction · Deep multi-task learning · Radiomics

1 Introduction

Head and Neck (H&N) cancers are among the most common cancers worldwide (5th by incidence) [1]. Outcome prediction is a crucial task for H&N cancer patients in clinical practice, as it provides prognostic information for clinicians to guide treatment planning at an early stage so as to improve the survival outcomes of patients. However, outcome prediction is a challenging task as survival outcomes are intrinsically driven by many influential factors, such as clinical demographics, treatment regimens, and disease

© The Author(s), under exclusive license to Springer Nature Switzerland AG 2023
V. Andrearczyk et al. (Eds.): HECKTOR 2022, LNCS 13626, pp. 135–143, 2023.
https://doi.org/10.1007/978-3-031-27420-6_14

physiology [9]. In addition, outcome prediction has to take into account incomplete survival data. Generally, survival data includes many right-censored samples, for which the exact time of events occurring (e.g., disease progression or recurrence) is unclear. For example, patients may be lost to follow-up, or the events are not observed due to limited follow-up time. In these cases, it is only known that the events did not occur in a certain period of time and the events might occur later. Outcome prediction models, to make the maximum use of existing information, need to build from both complete (uncensored) and incomplete (censored) samples.

The HEad and neCK TumOR segmentation and outcome prediction challenge (HECKTOR 2022) invites the research community to develop algorithms for tumor segmentation and outcome prediction in H&N cancers using pretreatment PET/CT [2, 3]. There are two tasks in this challenge: Task 1 - segmentation of primary tumors and lymph nodes in PET/CT images, and Task 2 - prediction of patient outcomes from PET/CT images and available clinical data (e.g., gender, age, etc.). Participating algorithms should perform fully automatic inference from entire PET/CT images without requiring bounding boxes. In this study, we focus on Task 2 and intend to improve outcome prediction. As our method also produces segmentation masks of tumor regions (primary tumors and lymph nodes) as an intermediate output, we also participated in Task 1 and report the related experimental results in this paper.

Radiomics, referring to the extraction and analysis of handcrafted quantitative features, is widely adopted in clinical practice for outcome prediction [4–6]. However, radiomics methods require manual segmentation of tumor regions for every patient, which is intractable and error-prone. To address this limitation, deep learning methods based on deep neural networks have been proposed, which perform end-to-end outcome prediction without requiring manual segmentation [7–9]. However, without segmentation masks, these methods take the whole image as input, which makes them difficult to focus on tumor regions and potentially unable to fully leverage the prognostic information in the tumor regions [12]. Recently, deep multi-task learning was introduced for joint outcome prediction and tumor segmentation [10–12]. Through jointly learning with tumor segmentation, deep multi-task models are implicitly guided to extract features related to tumor regions, which relieves the above-mentioned limitation of focusing on tumor regions. In addition, as mentioned above, bounding boxes are not given in this challenge. Under this circumstance, a deep multi-task model that performs both outcome prediction and tumor segmentation could be an optimal solution as it removes the requirement for additional tumor detection.

In our previous study, we proposed a Deep Multi-task Survival model (DeepMTS) for joint outcome prediction and tumor segmentation, and it has shown promising performance in H&N cancers [11] and Nasopharyngeal Carcinoma (NPC) [12]. However, although the DeepMTS has been shown to have the ability to focus on tumor regions by deep multi-task learning [12], we found it still cannot fully leverage the prognostic information in tumor regions. In this study, we extend the DeepMTS to a radiomics-enhanced deep multi-task framework, where radiomics features extracted from DeepMTS-segmented tumor regions were incorporated as an enhancement. We demonstrated that this extension improved DeepMTS by a large margin.

2 Materials and Methods

2.1 Patients

The organizers of HECKTOR challenge provided a training set of 524 patients acquired from 7 centers, including CHUM (n = 56), CHUP (n = 72), CHUS (n = 72), CHUV (n = 53), MDA (n = 198), HGJ (n = 55) and HMR (n = 18). A testing set of 359 patients was provided for evaluation, in which 200 patients are from a center present in the training set (MDA) while the other 159 patients are from two centers absent from the training set (USZ and CHB). Note that the ground-truth labels of the testing set were not released to the public and we did not use any external dataset in this challenge. All patients were histologically confirmed with oropharyngeal H&N cancer and received radiotherapy and/or chemotherapy. Each patient underwent FDG-PET/CT before treatment and was recorded with clinical indicators including age, gender, weight, treatment regimens, Human Papilloma Virus (HPV) status, performance status, and tobacco/alcohol consumption. For Task 1 (tumor segmentation), primary Gross Tumor Volume (GTVt) and nodal Gross Tumor Volumes (GTVn) were annotated by experts and were regarded as ground-truth labels. For Task 2 (outcome prediction), Recurrence-Free Survival (RFS), including time-to-event in days and event status, was provided as ground-truth labels. The provided event status is a binary indicator, where 1 indicates patients with observed disease recurrence and 0 indicates censored patients. Since some patients did not exhibit complete responses to the treatment, only 489 and 339 patients in the training and testing sets have RFS labels. Therefore, our method was trained and validated using 5-fold cross-validation with 489 patients in the training set, and then was tested with 359 and 339 patients in the testing set for Task 1 and Task 2.

2.2 Radiomics-Enhanced Deep Multi-task Framework

We propose a radiomics-enhanced deep multi-task framework that performs outcome prediction with PET/CT images and available clinical indicators. As shown in Fig. 1, the proposed framework adopts our recently proposed DeepMTS [12] as the backbone and then incorporates an automatic radiomics module as an enhancement. Specifically, DeepMTS is a deep multi-task model that can jointly predict the survival risk scores of patients and the segmentation masks of tumor regions (detailed in Sect. 2.3). With the segmentation masks derived from the DeepMTS, a radiomics module is used to extract discriminative features from the predicted tumor regions of PET/CT images (detailed in Sect. 2.4). We refer to this radiomics module as automatic radiomics, which differentiates it from traditional radiomics that relies on manual segmentation. Finally, the predicted survival risk scores, the discriminative radiomics features, and the clinical indicators are combined to build a Cox Proportional Hazard (CoxPH) model [13] for final outcome prediction.

2.3 Deep Multi-task Survival Model (DeepMTS)

Figure 2 shows the overall architecture of the DeepMTS [12], which consists of a U-net [14] based segmentation backbone, followed by a DenseNet [15] based cascaded survival

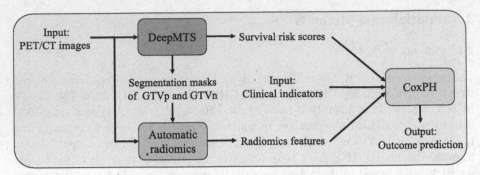

Fig. 1. The workflow of our radiomics-enhanced deep multi-task framework.

network (CSN). The segmentation backbone is hard-shared by tumor segmentation and outcome prediction tasks, which implicitly guides the model to extract features related to tumor regions. The outputs of the segmentation backbone are fed into the CSN as a supplementary input, which provides the CSN with global tumor information (e.g., tumor size, shape, and locations). Deep features are derived from both the segmentation backbone and the CSN, and then are used to predict survival risk scores through several fully-connected layers.

The DeepMTS is trained in an end-to-end manner to minimize a combined loss L of a segmentation loss L_{seg} and a survival prediction loss L_{surv}:

$$L = L_{seg} + L_{surv} \tag{1}$$

The L_{seg} is the Dice loss [16] between the predicted probability maps and the ground-truth labels, while the Cox negative logarithm partial likelihood loss [17] is used as the L_{surv} to handle right-censored survival data:

$$L_{surv} = -\frac{1}{N_{E=1}} \sum_{i:E_i=1} \left(h_i - \log \sum_{j \in \mathcal{H}(T_i)} e^{h_j} \right) \tag{2}$$

where h is the predicted risk scores, E is the event status, T is the time of RFS (for $E=1$) or the time of patient censored (for $E=0$), $N_{E=1}$ is the number of patients with disease recurrence, and $\mathcal{H}(T_i)$ is a set of patients whose T is no less than T_i.

DeepMTS has two advantages when compared to employing two separate segmentation and outcome prediction models: (i) The DeepMTS can produce more accurate survival risk scores than single-task survival models [12]; (ii) Through jointly learning with outcome prediction, the segmentation target might become closer to the regions providing prognostic information [21], which will potentially improve the performance of downstream automatic radiomics. Further discussion and related ablation analysis can be found in our previous work [12].

2.4   Automatic Radiomics

For automatic radiomics, handcrafted features are extracted from the predicted tumor regions of PET/CT images by Pyradiomics [18]. The predicted GTVp and GTVn masks

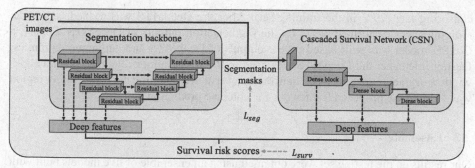

Fig. 2. The overall architecture of DeepMTS. More details can be found in [12].

are merged into a single mask (1 for GTVp and GTVn and otherwise 0) for feature extraction. The extracted handcrafted features include the features from First Order Statistics (FOS), Neighboring Grey Tone Difference Matrix (NGTDM), Grey-Level Run Length Matrix (GLRLM), Grey-Level Size Zone Matrix (GLSZM), Grey-Level Cooccurrence Matrix (GLCM), and 3D shape-based features. In addition to the original PET/CT, eight wavelet decompositions of PET/CT are also used, resulting in a total of 1689 radiomics features. All radiomics features are standardized using Z-score normalization and then are reduced to seven discriminative radiomics features through Least Absolute Shrinkage and Selection Operator (LASSO) regression.

2.5 Image Preprocessing

We resampled images into isotropic voxels of unit dimension to ensure comparability, where 1 voxel corresponds to 1 mm^3. Trilinear and nearest neighbor interpolations were used for PET/CT images and segmentation masks, respectively. Based on the provided segmentation ground-truth labels, each image in the training set was cropped to $160 \times 160 \times 160$ voxels with the tumor center aligning with the image center. However, as no bounding box was provided, each image in the testing set was cropped to $240 \times 240 \times 240$ voxels to ensure the entire head and neck region can be included. For intensity normalization, PET images were standardized using Z-score normalization, while CT images were clipped to $[-1024, 1024]$ and then mapped to $[-1, 1]$.

2.6 Training and Inference

We first implemented the DeepMTS using Keras with a Tensorflow backend on two 12 GB GeForce GTX Titan X GPUs. Specifically, we used an Adam optimizer with a batch size of 8 to train the DeepMTS for a total of 10,000 iterations. The learning rate was set as 1e−4 initially and then was reset to 5e−5 and 1e−5 at the 2,500th and 5,000th training iteration. Data augmentation was applied to the PET/CT images in real-time during training to minimize overfitting. The used data augmentation techniques include random affine transformations and random cropping to $112 \times 112 \times 112$ voxels. Moreover, we sampled an equal number of censored and uncensored samples during the data augmentation process to minimize the problem introduced by the imbalanced censoring rate (79% in the training set). After the DeepMTS was trained, automatic radiomics was performed based on the segmentation outputs of DeepMTS. Finally, a CoxPH model was built based on the outputs of DeepMTS and automatic radiomics. During inference, the DeepMTS was used to locate the tumor region through a slide-window segmentation process. Then, the image patch containing the whole tumor region was fed into the proposed framework for outcome prediction.

2.7 Ensemble

Our results on the testing set were obtained using an ensemble of five models built with 5-fold cross-validation. For outcome prediction (Task 2), the testing results of the five models were first standardized by Z-score normalization and then averaged together to obtain the final testing results. For tumor segmentation (Task 1), we tried two ensemble approaches: (i) The testing results of the five models were first averaged together and then thresholded at 0.5 to obtain the final testing results, and (ii) The testing results of the five models were first thresholded at 0.5 and the final testing results were obtained through majority voting.

2.8 Evaluation Metrics

Tumor segmentation (Task 1) is evaluated using aggregated Dice Similarity Coefficient (DSC) [19]. The DSC metric measures the similarity between the predicted and ground-truth segmentation masks, which is computed separately for GTVp and GTVn. Outcome prediction (Task 2) is evaluated using concordance index (C-index) [20]. The C-index metric measures the consistency between the predicted survival scores and the ground-truth survival outcomes.

3 Results and Discussion

3.1 Outcome Prediction

Our experimental results of outcome prediction are reported in Table 1. For 5-fold cross-validation, the average results in five folds are reported along with the range in parentheses. We first evaluated the performance of using clinical indicators, automatic radiomics, and DeepMTS, separately. Then, we combined them in our radiomics-enhanced deep multi-task framework. The results of traditional radiomics (based on ground-truth manual segmentation) are also reported in Table 1 for comparison. We found that the DeepMTS achieved higher validation C-index than clinical indicators and radiomics methods, which is consistent with the results reported in our previous study [12]. Nevertheless, the DeepMTS still can be further improved by embedding clinical indicators and automatic radiomics. This demonstrates that, although the DeepMTS has been shown to have the ability to focus on tumor regions through deep multi-task learning [12], it still cannot fully leverage the prognostic information in the tumor regions and the automatic radiomics could be an effective enhancement. In addition, we found that automatic radiomics achieved higher validation C-index than traditional radiomics, which suggests that the provided ground-truth tumor regions might not be the optimal regions for prognosis. As we mentioned in Sect. 2.3, the regions predicted by DeepMTS may provide more prognostic information due to joint learning with outcome prediction.

During the challenge, only a maximum of 3 testing submissions were allowed. We, therefore, submitted the methods with the top-3 validation C-index. The highest testing C-index was achieved by combining DeepMTS, clinical indicators, and automatic radiomics, which made us place the 2nd on the leaderboard with a negligible difference from the 1st place (0.00068 lower in C-index: 0.68084 vs 0.68152).

Table 1. The C-index results on the 5-fold cross-validation and testing set.

Method	5-fold cross-validation	Testing set
Clinical indicators	0.662 (0.613–0.689)	/
Traditional radiomics	0.681 (0.625–0.732)	/
Automatic radiomics	0.688 (0.644–0.735)	/
DeepMTS	0.705 (0.663–0.740)	0.644
DeepMTS + Clinical indicators	0.731 (0.702–0.773)	0.649
DeepMTS + Clinical indicators + Automatic radiomics	0.768 (0.723–0.786)	0.681

3.2 Tumor Segmentation

We also submitted the segmentation masks predicted by DeepMTS for Task 1. The testing results of using two different ensemble approaches (averaging and majority voting) are shown in Table 2. We found that averaging the outputs of five models (built with 5-fold cross-validation) achieved higher testing DSC than majority voting. We also found that the DeepMTS achieved higher testing DSC on GTVp than GTVn. This is because the DeepMTS was developed for primary tumor segmentation and we did not specially optimize it for lymph node segmentation. It should be noted that we focus on outcome prediction and the segmentation performance did not degrade our framework's performance in the outcome prediction. Instead, the predicted segmentation masks potentially improved the performance of radiomics (Table 1).

Table 2. The DSC results on the testing set.

Ensemble approach	GTVp	GTVn	Mean
Averaging	0.761	0.659	0.710
Majority voting	0.760	0.658	0.709

4   Conclusion and Limitations

In this study, we have outlined a radiomics-enhanced deep multi-task framework for outcome prediction in the context of HECKTOR 2022. With our recently proposed DeepMTS as the backbone, we incorporate automatic radiomics as an enhancement to further leverage the prognostic information available in the tumor regions. Our deep multi-task framework achieved a competitive result in the testing set of HECKTOR 2022, which made us place the 2nd in the leaderboard and only have a negligible difference (0.00068 lower in C-index) from the 1st place. Nevertheless, this study still has some limitations, and we suggest better performance potentially could be obtained by addressing the following:

Firstly, the balance between the losses of segmentation task and outcome prediction task was not fully explored. For simplicity, the segmentation loss L_{seg} and the survival prediction loss L_{surv} were directly summed without using any weighting parameters (equal weight was assigned to L_{surv} and L_{seg}), which might lead to sub-optimal performance. Secondly, the DeepMTS relies on an early fusion strategy where CT and PET images are concatenated as a 2-channel input. Other fusion strategies, such as intermediate or late fusion, could be further explored and this might enable better performance. Finally, due to limited time and computing resources, we performed 5-fold cross-validation to build the final model ensemble. However, leave-one-center-out validation might be a better alternative as this enforces models to learn prognostic information across different centers, which is helpful to build a robust ensemble.

References

1. Parkin, D.M., Bray, F., Ferlay, J., Pisani, P.: Global cancer statistics, 2002. CA: Cancer J. Clin. **55**(2), 74–108 (2005)
2. Andrearczyk, V., et al.: Overview of the HECKTOR challenge at MICCAI 2021: automatic head and neck tumor segmentation and outcome prediction in PET/CT images. In: Andrearczyk, V., Oreiller, V., Hatt, M., Depeursinge, A. (eds.) HECKTOR 2021. LNCS, vol. 13209, pp. 1–37. Springer, Cham. https://doi.org/10.1007/978-3-030-98253-9_1
3. Oreiller, V., Andrearczyk, V., Jreige, M., Boughdad, S., Elhalawani, H., Castelli, J., et al.: Head and neck tumor segmentation in PET/CT: the HECKTOR challenge. Med. Image Anal. **77**, 102336 (2022)
4. Vallieres, M., Kay-Rivest, E., Perrin, L.J., Liem, X., Furstoss, C., Aerts, H.J., et al.: Radiomics strategies for risk assessment of tumour failure in head-and-neck cancer. Sci. Rep. **7**(1), 10117 (2017)
5. Bogowicz, M., Riesterer, O., Stark, L.S., Studer, G., Unkelbach, J., et al.: Comparison of PET and CT radiomics for prediction of local tumor control in head and neck squamous cell carcinoma. Acta Oncol. **56**(11), 1531–1536 (2017)
6. Castelli, J., Depeursinge, A., Ndoh, V., Prior, J.O., Ozsahin, M., et al.: A PET-based nomogram for oropharyngeal cancers. Eur. J. Cancer **75**, 222–230 (2017)
7. Gu, B., Meng, M., Bi, L., Kim, J., Feng, D.D., Song, S.: Prediction of 5-year progression-free survival in advanced nasopharyngeal carcinoma with pretreatment PET/CT using multi-modality deep learning-based radiomics. Front. Oncol. **12**, 899351 (2022)
8. Saeed, N., Al Majzoub, R., Sobirov, I., Yaqub, M.: An ensemble approach for patient prognosis of head and neck tumor using multimodal data. In: Andrearczyk, V., Oreiller, V., Hatt, M.,

Depeursinge, A. (eds.) HECKTOR 2021. LNCS, vol. 13209, pp. 278–286. Springer, Cham (2022)

9. Naser, M.A., et al.: Progression free survival prediction for head and neck cancer using deep learning based on clinical and PET-CT imaging data. In: Andrearczyk, V., Oreiller, V., Hatt, M., Depeursinge, A. (eds.) HECKTOR 2021. LNCS, vol. 13209, pp. 287–299. Springer, Cham (2022)

10. Andrearczyk, V., et al.: Multi-task deep segmentation and radiomics for automatic prognosis in head and neck cancer. In: Rekik, I., Adeli, E., Park, S.H., Schnabel, J. (eds.) PRIME 2021. LNCS, vol. 12928, pp. 147–156. Springer, Cham (2021)

11. Meng, M., Peng, Y., Bi, L., Kim, J.: Multi-task deep learning for joint tumor segmentation and outcome prediction in head and neck cancer. In: Andrearczyk, V., Oreiller, V., Hatt, M., Depeursinge, A. (eds.) HECKTOR 2021. LNCS, vol. 13209, pp. 160–167. Springer, Cham (2022)

12. Meng, M., Gu, B., Bi, L., Song, S., et al.: DeepMTS: deep multi-task learning for survival prediction in patients with advanced nasopharyngeal carcinoma using pretreatment PET/CT. IEEE J. Biomed. Health Inform. **26**(9), 4497–4507 (2022)

13. Cox, D.R.: Regression models and life-tables. J. Roy. Stat. Soc.: Ser. B (Methodol.) **34**(2), 187–202 (1972)

14. Çiçek, Ö., Abdulkadir, A., Lienkamp, S.S., Brox, T., Ronneberger, O.: 3D U-Net: learning dense volumetric segmentation from sparse annotation. In: Ourselin, S., Joskowicz, L., Sabuncu, M.R., Unal, G., Wells, W. (eds.) MICCAI 2016. LNCS, vol. 9901, pp. 424–432. Springer, Cham (2016). https://doi.org/10.1007/978-3-319-46723-8_49

15. Huang, G., Liu, Z., Van Der Maaten, L., Weinberger, K.Q.: Densely connected convolutional networks. In: Proceedings of the IEEE Conference on Computer Vision and Pattern Recognition (2017)

16. Milletari, F., Navab, N., Ahmadi, S.A.: V-Net: fully convolutional neural networks for volumetric medical image segmentation. In: 2016 Fourth International Conference on 3D Vision (3DV). IEEE (2016)

17. Katzman, J.L., Shaham, U., Cloninger, A., Bates, J., Jiang, T., Kluger, Y.: DeepSurv: personalized treatment recommender system using a Cox proportional hazards deep neural network. BMC Med. Res. Methodol. **18**(1), 24 (2018)

18. Van Griethuysen, J.J., Fedorov, A., Parmar, C., Hosny, A., et al.: Computational radiomics system to decode the radiographic phenotype. Can. Res. **77**(21), e104–e107 (2017)

19. Kumar, N., et al.: A dataset and a technique for generalized nuclear segmentation for computational pathology. IEEE Trans. Med. Imaging **36**(7), 1550–1560 (2017)

20. Harrell Jr., F.E., Lee, K.L., Mark, D.B.: Multivariable prognostic models: issues in developing models, evaluating assumptions and adequacy, and measuring and reducing errors. Stat. Med. 15(4), 361–387 (1996)

21. Liu, J., et al.: A cascaded deep convolutional neural network for joint segmentation and genotype prediction of brainstem gliomas. IEEE Trans. Biomed. Eng. **65**(9), 1943–1952 (2018)

Recurrence-Free Survival Prediction Under the Guidance of Automatic Gross Tumor Volume Segmentation for Head and Neck Cancers

Kai Wang(iD), Yunxiang Li(iD), Michael Dohopolski(iD), Tao Peng(iD), Weiguo Lu(iD),
You Zhang(iD), and Jing Wang(✉)(iD)

Department of Radiation Oncology,
University of Texas Southwestern Medical Center, Dallas, USA
Jing.Wang@UTSouthwestern.edu
https://labs.utsouthwestern.edu/advanced-imaging-and-informatics-
radiation-therapy-airt-lab

Abstract. For Head and Neck Cancers (HNC) patient management, automatic gross tumor volume (GTV) segmentation and accurate pre-treatment cancer recurrence prediction are of great importance to assist physicians in designing personalized management plans, which have the potential to improve the treatment outcome and quality of life for HNC patients. In this paper, we developed an automated primary tumor (GTVp) and lymph nodes (GTVn) segmentation method based on combined pre-treatment positron emission tomography/computed tomography (PET/CT) scans of HNC patients. We extracted radiomics features from the segmented tumor volume and constructed a multi-modality tumor recurrence-free survival (RFS) prediction model, which fused the prediction results from separate CT radiomics, PET radiomics, and clinical models. We performed 5-fold cross-validation to train and evaluate our methods on the MICCAI 2022 HEad and neCK TumOR segmentation and outcome prediction challenge (HECKTOR) dataset. The ensemble prediction results on the testing cohort achieved aggregated Dice scores of 0.77 and 0.73 for GTVp and GTVn segmentation, respectively, and a C-index value of 0.67 for RFS prediction. The code is publicly available (https://github.com/wangkaiwan/HECKTOR-2022-AIRT). Our team's name is AIRT.

Keywords: Head and neck cancer · Automatic segmentation · Recurrence-free survival prediction

---

K. Wang and Y. Li—Equal contribution.

1   Introduction

Head and Neck cancer (HNC) is one of the most common cancers worldwide [1]. With the development of HNC radiotherapy and chemo-radiotherapy, HNC patients can be successfully treated in many cases. However, even treated with curative intent, there are still more than 15% patients which will experience cancer recurrence, and locoregional failures occur in up to 40% of patients in the first two years after the treatment [2]. Therefore, a strategy that can accurately identify HNC patients at high risk of recurrence at diagnosis would be helpful in assisting physicians in making personalized treatment plans, which have the potential to improve patient treatment outcomes.

Positron Emission Tomography (PET) and Computed Tomography (CT) imaging (PET/CT) play an important role in the management of HNC patients [3]. At initial staging, PET/CT plays the central role in characterizing local, regional and distant disease, while during therapy and after therapy, it is routinely used to assess the treatment response and detect recurrence and/or metastases. Recently, using PET/CT scans of HNC patients, several radiomics and deep learning studies were proposed to better predict the treatment outcomes, including locoregional recurrence, distant recurrence, and recurrence-free survival (RFS) [4–8]. Besides, as deep learning-based image segmentation has achieved comparable performance to humans in many tasks, several organs at risk (OAR) and gross tumor volume (GTV) automatic segmentation methods were proposed to help in the management of HNC patients [9–11]. These segmentation methods not only have the potential to ease the workload of healthcare workers but also assist in the workflow of radiomics or deep learning-based treatment outcome prediction models, which can heavily rely on the accuracy of the region of interest (ROI) delineation.

Although the promising performance was reported in some of the automatic segmentation and outcome prediction studies, their reliability, robustness, and reproducibility are still of concern for clinical translation. Aiming to further study the automatic segmentation and RFS prediction method for HNC tumors and exam their cross-institutional performance objectively, the MICCAI 2022 Head and NeCK TumOR segmentation and outcome prediction challenge (HECKTOR) offered researchers the opportunity to build and evaluate their auto-segmentation and RFS prediction models with a very large dataset collected from multiple institutions [12,13]. In the present work, we built an auto-segmentation framework for precise segmentation of HNC primary tumor (GTVp) and lymph nodes (GTVn) with HECKTOR 2022 dataset, and we introduced a multi-modality HNC RFS prediction framework that ensembles the predicted risk scores from separate clinical feature model, PET radiomics model, and CT radiomics model. The deep auto-segmentation model is used to identify the ROIs for radiomics feature extraction.

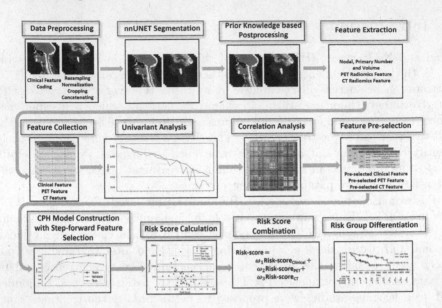

Fig. 1. A schematic overview of our proposed automatic gross tumor volume segmentation guided cancer recurrence free survival prediction framework.

2   Material and Methods

2.1   Head and NeCK TumOR 2022 (HECKTOR 2022) Dataset

The dataset provided in HECKTOR 2022 consisted of PET/CT scans, clinical features, and follow-up information of 845 HNC patients from 7 centers. The total number of training cohorts is 524 and 489 for tasks 1 (GTVp and GTVn segmentation) and 2 (RFS prediction), respectively. Training and test cohorts are representative of the distribution of the real-world population of patients accepted for initial staging of oropharyngeal cancer (with around 21% of recurrence and a median RFS of 14 months in the training set). For the PET images, computation of the Standardized Uptake Value (SUV) is already done by the challenge organizer. Missing patient weight was estimated as 75 Kg to compute SUV for a small subset of patients (8 of 845). All the PET/CT files conversed from the DICOM file format to NIfTI format. More detailed description of the dataset can be found [12,13]. The provided clinical features are gender, age, tobacco and alcohol consumption, performance status (Zubrod), human papillomavirus (HPV) status, treatment (surgery and/or chemotherapy in addition to the radiotherapy that all patients underwent). There is some clinical information missing (tobacco, alcohol, performance status, HPV status), and we encoded the missing feature with value 0, known as negative status as −1. Five-fold cross-validation strategy was used to train and validate our method for both two tasks. And we fused the prediction results of 5-fold models for the testing cohort to generate the ensemble result for the challenge submission.

2.2 Overall Architecture

The workflow of our proposed method is shown in Fig. 1. After PET/CT image processing, we trained a nnUNET model with concatenated PET/CT image as the input, and the outputs are GTVp and GTVn masks. Then we calculated the predicted GTVp and GTVn number and volume as additional clinical features. Within the predicted GTVp volume, we extracted PET, and CT radiomics feature separately. After univariate analysis and feature correlation analysis, we removed the low-predictive features and redundant features. The remaining features would go through a step-forward feature selection with Cox Proportional Hazards (CPH) model, and the best model and feature set was determined by the validation performance. Then the risk score of testing patients can be calculated. We trained the clinical feature model, CT radiomics model, and PET radiomics model separately. Their predicted risk score for testing data was fused together as the multi-modality RFS model prediction results to predict patients' RFS and identify high- or low-risk patients.

2.3 Gross Tumor Volume Segmentation (Task 1)

nnUNet for Segmentation. We used the nnUNet [14] 3D full-resolution pipeline as the basis segmentation backbone. Considering each patient image had a different voxel size, the original PET/CT images were resampled to 2 mm × 2 mm × 2 mm pixel spacing with trilinear interpolation. The labels were resampled to the same size spacing. Besides, the PET/CT images were cropped into 2 × 128 × 128 × 128 patches with overlaps. Each image was standardized using z-scores. To avoid overfitting, image augmentation was used with random rotation in all directions between [−30, 30] degrees and random scaling between [0.7, 1.4]. All 524 patients were randomly divided into five groups for the five-fold cross-validation, each time leaving out one group of patients for validation and the others for training.

The training protocol is listed in Table 1.

Table 1. Training protocols used for training and optimization of our method.

Parameter	Setting
Patch size	128 × 128 × 128
Batch size	2
weight_decay	3e−05
Optimizer SGD	SGD
Initial learning rate	0.01
Learning rate decay schedule	polyLR
Nesterov momentum	0.99
Epoch	1000

Post-processing. It is known a priori from anatomical knowledge that the distances between the primary tumor (GTVp) and lymph nodes (GTVn) are often under certain thresholds. In order to remove unreasonable segmentation results, we first calculate the center-of-mass coordinates of GTVp and GTVn. Assume the center-of-mass coordinates of GTVp and GTVn are (x_p, y_p, z_p) and (x_n, y_n, z_n), separately. The distance D_{pn} between GTVp and GTVn can be calculate via Eq. (1).

$$D_{pn} = \sqrt{(x_p - x_n)^2 + (y_p - y_n)^2 + (z_p - z_n)^2} \tag{1}$$

Based on the clinician's empirical judgment, we set $D_{max} = 150$ mm as the threshold value for distance. Then, we remove the lymph nodes whose distance is greater than D_{max}.

2.4   Recurrence-Free Survival Prediction (Task 2)

With the nnUNET predicted the GTVp and GTVn volumes on the resampled PET/CT image, we extract PET and CT radiomics feature separately from the automatic segmented GTVp and GTVn volumes for both training and testing datasets with the pyradiomics package [15]. Default feature extraction settings were used for both PET and CT feature extraction. As the number and volume of GTVp and GTVn are reported to be predictive for HNC treatment outcome prediction, and they are not as complex features as radiomics features, we added them to clinical feature set. For the missing value in clinical feature, comprising 293 tobacco consumption, 326 alcohol consumption, 268 performance status, and 167 missing HPV status, we coded the positive status (1 in the original file) to 1, negative status (0 in the original file) to −1, and missing value to 0.

For training and validating our method with training cohort data, we followed the same splitting as what we did for training the nnUNET. For each training fold in the 5-fold cross-validation procedure, we first performed 100 times internal 5-fold cross-validation on the training fold for clinical feature, CT radiomcis feature, PET radiomics feature pre-selection separately. The partition of the internal training and validation is random each time. Univariant cox regression models were built with each feature for RFS prediction. The average C-index of univariant RFS prediction models validation data was recorded, and the order of features was sorted based on the validation performance. To avoid the low-predictive-ability features, the feature that has a lower than 0.5 average C-index was removed from the feature set. To reduce redundancy in the feature sets for different prediction targets, we performed Pearson correlation analysis for all the features. A feature that has an absolute correlation coefficient higher than 0.9 to any of its previous features was removed from the feature set.

Then, we performed multivariate CPH regression with the step forward feature selection strategy to further select predictive features and construct the RFS prediction models. C-index was the criteria for the step forward feature selection in this step, 5 was set as the maximum number of selected clinical features, 10 was set as the maximum number of selected radiomics features. Another internal 100 times 5-fold cross-validation was conducted to mitigate the impact of random patient partition. The risk scores of training samples and validation samples in each time of 5-fold cross-validation were recorded. The combination of the 5-fold risk scores was used as the final survival prediction on the training cohort of each single-modality model, and the C-index was used to evaluate our models, which quantifies the model's ability to provide an accurate ranking of the survival times based on the computed individual risk scores, generalizing the Area Under the ROC Curve (AUC). We then calculated the average risk scores of the three single-modality models for testing patients and used them as our final prediction (multi-modality RFS prediction model results). Of note, for patients who were predicted to have no GTVp volume (4 in the training cohort, 2 in the testing cohort), their final prediction results are the same as clinical feature-based models, as they don't have GTVp radiomics features. We also used the multi-modality risk score to differentiate between high- and low-risk patient groups. The threshold was set as 0. We performed Kaplan-Meier analysis and log-rank test to evaluate the RFS difference of the identified patient groups, a P-value $\leqslant$ 0.05 was considered significant. Lifelines python package was used to construct the model and perform statistic analysis.

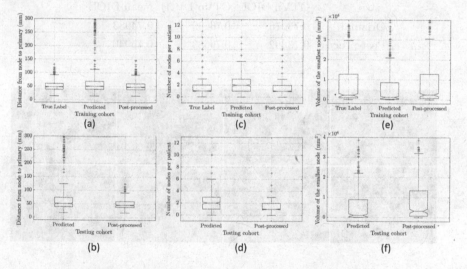

Fig. 2. Node properties of label, predicted and post-processed data on training (a, c, e) and testing (b, d, f) cohorts. The analyzed properties include distance from nodes to center of primary tumor for each patients (a, b), number of nodes per patients (c, d), and volume of the smallest node of each patient (e, f).

3   Results

3.1   Gross Tumor Volume Segmentation (Task 1)

Due to the fact that the PET/CT images contain most of the patient body while the lymph nodes are only distributed in small regions, some regions of the patient bodies with high PET image intensity were misidentified as lymph nodes by our model without post-processing, which is illustrated in Fig. 3. With post-processing, our method is more capable of identifying the segmentation result of other parts as unreliable results and excluding it. Besides, as shown in Fig. 2, the data distribution of several metrics becomes closer to the manual label, including the number of nodes, volume of the smallest node, and distance from the node to the primary. As shown in Table 2, we quantitatively compare the segmentation results before and after post-processing, and we can see that our DICE has some improvement after processing.

Table 2. Nodal and primary tumor segmentation Dice similarity coefficients before and after post-processing on testing cohort.

	GTVn DICE	GTVp DICE	mean DICE
Original	0.73276	0.76689	0.74983
Post-process	0.73392	0.76689	0.75040

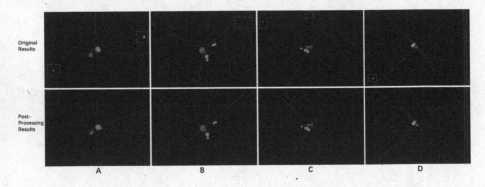

Fig. 3. Visualization comparison of the post-processed results and the original results. Obvious error areas are marked with red boxes. (Color figure online)

3.2   Recurrence-Free Survival Prediction (Task 2)

The RFS prediction performance evaluated with C-index on training and testing cohort is shown in Fig. 3. The Kaplan-Meier analysis and log-rank test results of the RFS difference of different patient risk groups identified by single-modality and multi-modality RFS models are shown in Fig. 4. To our surprise, although there are a lot of missing data in the clinical feature set, the clinical feature-based model is still very predictive (C-index = 0.68 on training cohort), and the top predictive clinical features are GTVp volume, Node volume, HPV status, and tobacco consumption. The CT-radiomics model is the most predictive single-modality model, and its C-index on the training cohort is 0.69. The fused multi-modality RFS prediction model has the best prediction performance in the training cohort, with a prediction C-index value of 0.72. On the testing cohort, our submitted multi-modality RFS risk score got a C-index value of 0.67, which ranked No. 3 on the challenge leaderboard.

Table 3. Single-modality and Multi-modality models performance for recurrence-free survival prediction on training and testing cohort.

Modality	Clinical	CT Radiomics	PET Radiomics	Multi-Modality
C-index (Training)	0.68253	0.68927	0.62535	0.72036
C-index (Testing)	—	—	—	0.67257

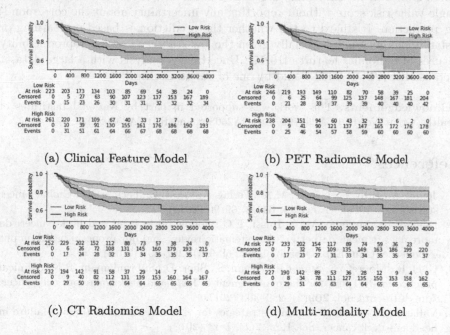

(a) Clinical Feature Model   (b) PET Radiomics Model

(c) CT Radiomics Model   (d) Multi-modality Model

Fig. 4. Kaplan-Meier analysis of risk scores from single- and multi-modality recurrence-free survival (RFS) models on training cohort. A risk score value of 0 was used to identify the high- and low-risk patient group. For all four models, the identified low-risk groups have significantly better RFS than the high-risk group, the $-log2(p)$ values of them are 11.84, 8.05, 15.89, and 16.98, respectively.

4   Discussion and Conclusion

In this paper, we presented a multi-modality HNC RFS prediction method guided by auto-segmentation GTV segmentation, which achieved a 0.67 C-index for RFS prediction and 0.75 DSC for GTV segmentation. Although the results got good rankings in the challenge, there are still some limitations to our work. Firstly, our segmentation-guided RFS prediction model is not optimized end-to-end, current separate training strategy might not yield the best performance. Secondly, as radiomics method heavily relies on accurate ROI delineation, it's important to evaluate the performance difference between the model built with features extracted from human expert delineation and features extracted from auto-segmented volume. Thirdly, in the current experiment setting we averaged the risk scores from different modality models to generate the risk score of the proposed multi-modality model, which is not optimal. Weighting the risk scores from different models based on their validation performance or optimizing these weighting factors via training can be a better way. Finally, as RFS prediction is a medical task, irresponsible prediction can be costly if a RFS prediction model is used in the clinic. However, our current RFS model can only generate a single-value risk score without reporting any uncertainty about the corresponding prediction. It's hard to tell whether the prediction is based on learning or just random guessing, especially given the low-stability and low-reproducibility natures of radiomics features [16,17]. Uncertainty analysis with a large dataset such as HECKTOR 2022 would be one of our future work.

Acknowledgements. This work was supported in part by National Institutes of Health (Grant No. R01CA251792, R01CA240808 and R01CA258987).

References

1. Jemal, A., Bray, F., Center, M.M., Ferlay, J., Ward, E., Forman, D.: Global cancer statistics. CA Cancer J. Clin. **61**(2), 69–90 (2011)
2. Denaro, N., Merlano, M.C., Russi, E.G.: Follow-up in head and neck cancer: do more does it mean do better? A systematic review and our proposal based on our experience. Clin. Exp. Otorhinolaryngol. **9**(4), 287–297 (2016)
3. Goel, R., Moore, W., Sumer, B., Khan, S., Sher, D., Subramaniam, R.M.: Clinical practice in PET/CT for the management of head and neck squamous cell cancer. Am. J. Roentgenol. **209**(2), 289–303 (2017)
4. Vallieres, M., et al.: Radiomics strategies for risk assessment of tumour failure in head-and-neck cancer. Sci. Rep. **7**(1), 1–14 (2017)
5. Diamant, A., Chatterjee, A., Vallières, M., Shenouda, G., Seuntjens, J.: Deep learning in head & neck cancer outcome prediction. Sci. Rep. **9**(1), 1–10 (2019)
6. Wang, K., et al.: A multi-objective radiomics model for the prediction of locoregional recurrence in head and neck squamous cell cancer. Med. Phys. **47**(10), 5392–5400 (2020)
7. Chen, L., et al.: Attention guided lymph node malignancy prediction in head and neck cancer. Int. J. Radiat. Oncol. Biol. Phys. **110**(4), 1171–1179 (2021)

8. Wang, R., et al.: Locoregional recurrence prediction in head and neck cancer based on multi-modality and multi-view feature expansion. Phys. Med. Biol. **67**(12), 125004 (2022)

9. van Rooij, W., Dahele, M., Brandao, H.R., Delaney, A.R., Slotman, B.J., Verbakel, W.F.: Deep learning-based delineation of head and neck organs at risk: geometric and dosimetric evaluation. Int. J. Radiat. Oncol. Biol. Phys. **104**(3), 677–684 (2019)

10. Zhu, W., et al.: AnatomyNet: deep learning for fast and fully automated whole-volume segmentation of head and neck anatomy. Med. Phys. **46**(2), 576–589 (2019)

11. Nikolov, S., et al.: Deep learning to achieve clinically applicable segmentation of head and neck anatomy for radiotherapy. arXiv preprint arXiv:1809.04430 (2018)

12. Oreiller, V., et al.: Head and neck tumor segmentation in PET/CT: the HECKTOR challenge. Med. Image Anal. **77**, 102336 (2022)

13. Andrearczyk, V., et al.: Overview of the HECKTOR challenge at MICCAI 2021: automatic head and neck tumor segmentation and outcome prediction in PET/CT images. In: Andrearczyk, V., Oreiller, V., Hatt, M., Depeursinge, A. (eds.) Head and Neck Tumor Segmentation and Outcome Prediction. HECKTOR 2021. LNCS, vol. 13209. Springer, Cham (2022). https://doi.org/10.1007/978-3-030-98253-9_1

14. Isensee, F., Jaeger, P.F., Kohl, S.A., Petersen, J., Maier-Hein, K.H.: nnU-Net: a self-configuring method for deep learning-based biomedical image segmentation. Nat. Methods **18**(2), 203–211 (2021)

15. Van Griethuysen, J.J., et al.: Computational radiomics system to decode the radiographic phenotype. Cancer Res. **77**(21), e104–e107 (2017)

16. Wang, K., Dohopolski, M., Zhang, Q., Sher, D., Wang, J.: Towards reliable head and neck cancers locoregional recurrence prediction using delta-radiomics and learning with rejection option (2022)

17. Rizzo, S., et al.: Radiomics: the facts and the challenges of image analysis. Eur. Radiol. Exp. **2**(1), 1–8 (2018)

Joint nnU-Net and Radiomics Approaches for Segmentation and Prognosis of Head and Neck Cancers with PET/CT Images

Hui Xu[1,2], Yihao Li[2], Wei Zhao[3], Gwenolé Quellec[2], Lijun Lu[1(✉)], and Mathieu Hatt[2]

[1] School of Biomedical Engineering, Southern Medical University, Guangzhou, China
ljlubme@gmail.com
[2] LaTIM, INSERM, UMR 1101, Univ Brest, Brest, France
[3] Department of Colorectal Surgery, Chinese Academy of Medical Sciences and Peking Union Medical College, Beijing, China

Abstract. Automatic segmentation of head and neck cancer (HNC) tumors and lymph nodes plays a crucial role in the optimization treatment strategy and prognosis analysis. This study aims to employ nnU-Net for automatic segmentation and radiomics for recurrence-free survival (RFS) prediction using pretreatment PET/CT images in a multi-center HNC cohort of 883 patients (524 patients for training, 359 for testing) provided within the context of the HECKTOR MICCAI challenge 2022. A bounding box of the extended oropharyngeal region was retrieved for each patient with fixed size of $224 \times 224 \times 224$ mm^3. Then the 3D nnU-Net architecture was adopted to carry out automatic segmentation of both primary tumor and lymph nodes. From the predicted segmentation mask, ten conventional features and 346 standardized radiomics features were extracted for each patient. Three prognostic models were constructed containing conventional and radiomics features alone, and their combinations by multivariate CoxPH modelling. The statistical harmonization method, ComBat, was explored towards reducing multicenter variations. Dice score and C-index were used as evaluation metrics for segmentation and prognosis task, respectively. For segmentation task, we achieved a mean dice score of 0.7 for primary tumor and lymph nodes. For recurrence-free survival prediction, conventional and radiomics models obtained C-index values of 0.66 and 0.65 in the test set, respectively, while the combined model did not improve the prognostic performance (0.65).

Keywords: Head and Neck cancer · PET/CT · nnU-Net · Radiomics

Team name: RokieLab.
H. Xu and Y. Li—Contributed equally to this work.

1 Introduction

Head and Neck cancer (HNC) is the fifth most common cancer worldwide [22]. Radiotherapy combined with cetuximab has been established as standard treatment [6]; however, 40% of patients still experience loco-regional failures in the first two years after the treatment [7]. Early prediction of prognosis response is crucial to tailor individualized treatment strategies for improving long-term survival of HNC patients. Positron Emission Tomography/Computed Tomography (PET/CT) imaging has been reported as a powerful tool in managing HNC patients including diagnosis, staging, design of the radiotherapy planning, and prognosis evaluation [9]. Besides conventional metrics (i.e., TNM stage, tumor volume, SUV), radiomics features have also been shown to provide potential value in clinical decision-making for HNC patients. For instance, Vallières, et al. demonstrated the potential of radiomics for assessing the risk of specific tumour outcomes using multiple stratification groups [24]. Bogowicz, et al. investigated the prognostic value of both PET and CT radiomics features, and reported that their combinations showed better discriminative power than individual modality alone for local tumor control modelling in HNC [4]. However both conventional features and radiomics approaches are heavily dependent on the accurate identification and segmentation of the primary tumor and lymph node regions, which have been mostly determined through manual or semi-automatic contouring in previous studies. However, this process is tedious, time consuming and suffers from intra- and inter-expert variability. Thus, a fully automatic segmentation method [21] would greatly assist in designing radiotherapy planning and early assessing prognosis via quantitative metrics, such as radiomics.

The HEad and neCK TumOR (HECKTOR) MICCAI challenge 2021 provided the opportunity for participants to develop and evaluate segmentation approaches of the primary tumor and prognosis (progression-free survival) prediction [1] with a dataset of about 400 patients from 6 centers. However, for a convincing validation and promote the transfer to clinical practice, both segmentation and prognosis models need to be validated on larger and multi-center cohorts. Moreover, expansion of radiomics analysis with lymph nodes features was reported with significant improvement in prognosis prediction compared to the analysis of primary tumor alone [5]. In this context, the challenge HECKTOR 2022 contains more data and extends the scope of investigation [3]. First, lymph nodes were added into the segmentation task. Second, a larger multi-center cohort was provided with a total of 883 HNC patients. Third, no bounding boxes were provided so challengers had to implement a fully automatic pipeline including lesions detection. The challenge of HECKTOR 2022 thus proposed two tasks: 1) fully automatic detection and segmentation of primary tumor Gross Tumor Volumes (GTVp) and lymph nodes (GTVn) and 2) prediction of Recurrence-Free Survival (RFS).

In this study, we adopted the 3D nnU-Net as our baseline model for the segmentation task since it demonstrated excellent results in biomedical images segmentation without manual intervention [12] and has shown good performance in HECKTOR 2021 [19]. Additionally, we propose a modified version using pseudo labelling technique followed by 3D nnU-Net. Based on the predicted segmentations of GTVp and GTVn, we trained Cox Proportional Hazards regression model (CoxPH) combined with conventional and/or radiomics features for predicting the risk of RFS. The workflow of this study is shown in Fig. 1.

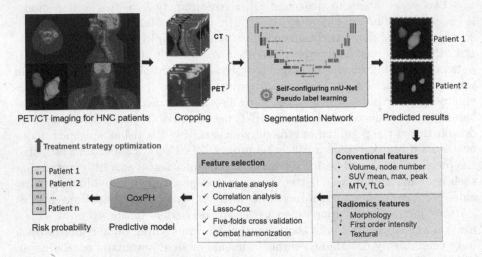

Fig. 1. The workflow of this study including images pre-processing, segmentation and prognosis prediction.

2   Method

2.1   Dataset Description

A total of 883 patients from 9 centers with histologically proven oropharyngeal HNC cancer were analyzed. A total of 524 patients from 7 centers was provided as a training set for Task 1. Of these, 489 patients were available for Task 2. The testing set contained 359 patients from 3 centers that were unseen in previous year's challenge [1]; specifically, 359 and 339 patients were used for Task 1 and Task 2, respectively. Of note, the center MDA provided data in both the training and testing sets. Details of dataset are provided in Table 1. In the training set, pre-treatment FDG PET/CT images, segmentation mask including primary tumor (GTVp) and lymph nodes (GTVn), clinical and prognosis information were provided for each patient. All the image dataset is in NIfTI format, and the segmentation mask has the same resolution as the CT images. Clinical data including center, gender, age and weight without missing values, as well as other variables such as HPV status, but with missing values. In the test set, only pre-treatment PET/CT images and clinical information were provided.

Table 1. The statistic and partitioning of multi-center HNC datasets.

No	Center	Devices	Task 1	Task 2	Cohort
1	CHUM	Discovery STE, GE Healthcare	56	56	Training
2	CHUP	Biography mCT 40 ToF, Siemens	72	44	Training
3	CHUS	Gemini GXL 16, Philips	72	72	Training
4	CHUV	Discovery D690 ToF, GE Healthcare	53	47	Training
5	MDA	Discovery HR, Discovery RX, Discovery ST, Discovery STE (GE, Healthcare)	198/200	197/200	Training/Test
6	HGJ	Discovery ST, GE Healthcare	55	55	Training
7	HMR	Discovery STE, GE Healthcare	18	18	Training
8	USZ	Discovery HR, Discovery RX, Discovery STE, Discovery LS, Discovery 690 (GE Healthcare)	101	101	Test
9	CHB	GE710, GE Healthcare	58	58	Test

2.2 Pre-processing

Since the covered regions of images are not consistent between patients, we adopted an automatic method [2] to retrieve head and neck regions. This method relies on anatomical information and PET intensity as a prior to find the brain region. A fixed size bounding box of $224 \times 224 \times 224$ mm^3 located three centimeters shift downward and forward from the lowest voxel of brain region was first determined. We evaluated the results by checking whether the GTVp and GTVn were fully contained in the bounding box according to the provided segmentation, where 509 out of 524 (97.1%) cases were correctly detected. In testing set, we only checked that the intensity range of PET and CT images was normal and not fully zero. 353 out of 356 (99.2%) cases were correctly detected. For the failure cases, a semi-automatic method by setting the center voxel was adopted to get the location of bounding box with same size. Once the bounding box location was confirmed, we cropped and resampled the original PET, CT and mask images to same scale of $224 \times 224 \times 224$ with the isotropic voxel size of $1 \times 1 \times 1$ mm^3 by linear interpolation. No further intensity normalization was performed on CT and PET images.

2.3 Task 1: Segmentation Prediction

Network Architecture. The architecture of the 3D nnU-Net is shown in Fig. 2. Before the launch of segmentation process, nnU-Net cropped the non-zero area of the PET and CT bounding box automatically. Then a patch-based sliding window technique was applied to the current cropped images, producing the

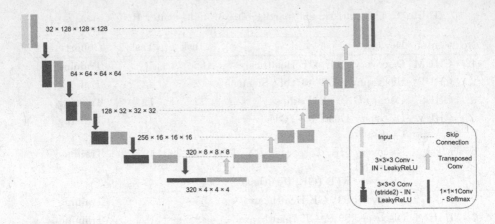

Fig. 2. The architecture of the 3D nnU-Net.

patches with size of $128 \times 128 \times 128$. These patches were input into the 3D nnU-Net. Two Conv-InstanceNorm-LeakyReLU blocks for down-sampling and up-sampling of the encoder and decoder was used in nnU-Net. Down-sampling is done by strided convolution, while up-sampling was done by transposed convolution. The architecture initially used 30 feature maps, which are doubled for each down-sampling operation in the encoder (up to 320 feature maps) and halved for each transpose convolution in the decoder. The end of the decoder has the same size as the input, followed by a $1 \times 1 \times 1$ convolution and a soft-max function.

Implementation Details. A combination of dice and cross-entropy loss is used to train our networks:

$$L_{total} = L_{dice} + L_{CE} \tag{1}$$

The dice loss formulation used here is adapted from the variant proposed in [13]. It is implemented as follows:

$$L_{dice} = -\frac{2}{|K|} \sum_{k \in K} \frac{\sum_{i \in I} u_i^k v_i^k}{\sum_{i \in I} u_i^k + \sum_{i \in I} v_i^k} \tag{2}$$

where u is the soft-max output of the network and v is the one hot encoding for the ground truth segmentation map. Both u and v have shape $I \times K$ with $i \in I$ being the pixels in the training patch/batch and $k \in K$ being the classes.

Adam optimizer was used with an initial learning rate of 3×10^{-4}. We trained for 1000 epochs, one epoch being defined as iteration over 250 mini batches. Batch size was set at 2. A five-fold cross-validation procedure was used, and 1000 epochs were trained per fold. Whenever the exponential moving average of the training losses did not improve by at least 5×10^{-3} within the last 30 epochs, the learning rate was reduced by factor 5. The training was terminated automatically if the exponential moving average of the validation losses did not improve by more than 5×10^{-3} within the last 60 epochs, but not before the learning rate was smaller than 10^{-6} [13].

During training, the following augmentation techniques were applied: random rotations, random scaling, random elastic deformations, gamma correction augmentation and mirroring. NVIDIA RTX A6000 48G was used in training. During inference, similar sliding window method was first used to generate patches: four $128 \times 128 \times 128$ voxel patches were processed and the predicted output probabilities were averaged in the overlapping regions.

2.4 Task 2: Prognosis Prediction

Conventional Features Development. Previous studies have reported the clinical prognostic values of conventional features in PET/CT imaging [23], thus a total of ten conventional features were firstly calculated using an in-house developed package based on the predicted segmentations of 3D nnU-Net. Conventional features included primary tumor volume, diameter, number of nodes, maximum, mean and peak standardized uptake value (SUVmax, SUVmean, SUVpeak), metabolic tumour volume based on the threshold of SUV2.5 and 40% SUVmax separately (MTV2.5, MTV40%), and total lesion glycolysis using corresponding MTV multiplied by SUVmean (TLG2.5, TLG40%). Specifically, these parameters related to SUV were calculated across both primary tumor and lymph nodes regions.

Radiomics Features Extraction. Radiomics features (shape, intensity, textures) were extracted from both PET and CT images separately using the open-source package of Standardized Environment for Radiomics Analysis (SERA) (https://github.com/ashrafinia/SERA), which conforms to the image biomarker standardisation initiative (IBSI) benchmark [27]. Based on the segmentation mask (with both primary tumor and lymph nodes) produced by the 3D nnU-Net, all default features were extracted using standard settings: isotropic voxel sizes of $2 \times 2 \times 2$ mm^3, fixed bin number (FBN) discretization with 64 bins. This resulted in a total of 346 radiomics features for each patient.

Cross-Validation Strategy. To evaluate the generalization performance of our prognostic models, we used a cross validation strategy. This step was used for feature selection and hyper-parameters adjustment in modelling. Since the provided testing set contained 200 patients from the center MDA, we randomly chose 97 patients (~20% of all training data) from the MDA center in the training set used as a separate fold namely fold 5, in order to select the prognostic model (developed by the training set) with relatively consistent performance in testing set. The remaining training cases (392 patients) were randomly split into four folds of 98 patients.

Model Construction. Three prognostic models were submitted which are based on 1) conventional features alone; 2) radiomics features alone combined with Combat harmonization strategy [14]; 3) conventional features combined with radiomics features without harmonization. For model 1, feature selection included univariate CoxPH model keeping features with the concordance index (C-index) [10] higher than 0.50, and Spearman rank correlation analysis to remove redundant features ($\rho_c < 0.60$). For model 2, Combat harmonization (non-parametric version) [14] was first applied for all radiomics features within the joint training and test sets using their center labels. Patient's gender, age and weight were used in the covariate matrix to preserve them in the harmonization. The least absolute shrinkage and selection operator Cox regression algorithm (Lasso-Cox) was adopted to identify the optimal features set. For model 3, the same features selection was used for radiomics features as for model 2, without Combat harmonization. Then the optimal radiomics and conventional features set were integrated as prognostic predictors. After feature selection, multivariate CoxPH model was constructed based on the entire training set to predict the survival risk (RFS) of HNC patients.

2.5 Evaluation Metrics

Aggregated Dice score [16] from both GTVp and GTVn was used as evaluation metric for segmentation task. The concordance index (C-index) between the predicted risk and the survival outcomes was used to measure the predictive performance in RFS prediction task.

Table 2. Segmentation results (Dice score) of two submitted models in the training and test sets.

Model	Training			Testing		
	GTVp	GTVn	mean	GTVp	GTVn	mean
nnU-Net (baseline)	0.86724	0.81576	0.84150	0.69997	0.70039	0.70018
nnU-Net with PLL	0.82551	0.75506	0.79028	0.70131	0.70100	0.70115

3 Results

3.1 Task 1: Segmentation Results

Table 2 provides the segmentation results (Dice score) of GTVp, GTVn and their averaged value on the training and test sets. The baseline segmentation model using 3D nnU-Net showed a mean Dice score of 0.70 across primary tumor and lymph nodes regions. Once baseline nnU-Net model was trained, We further adopted the pseudo-labeling learning (PPL) technique to enhance the performance of segmentation model. PPL technique consists in adding predicted test data to the training data to optimize parameters learning [17]. Thus, 3D nnU-Net with PPL was combined to retrain the nnU-Net model which integrated original training set and the predicted testing set as an updated training set. The updated model led to the same mean dice of 0.70 in the testing set. Examples of good and poor segmentation are displayed in Fig. 3.

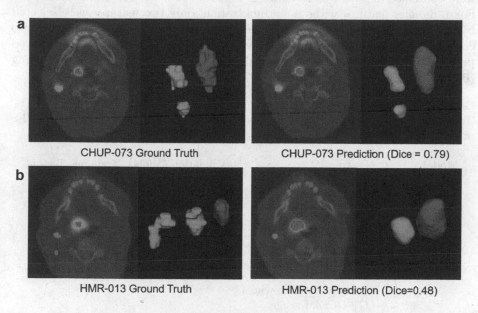

Fig. 3. An example of two patients (CHUP-073, HMR-013) for comparing the ground truth contours with the predicted mask by 3D nnU-Net.

Table 3. The individual performance of ten conventional features in five-folds cross validation.

Features	Volume	Diameter	Number of nodes	SUVmax	SUVmean
Mean C-index	0.5736	0.5611	0.6029	0.5924	0.5926
Features	SUVpeak	MTV2.5	MTV40%	TLG2.5	TLG40%
Mean C-index	0.6028	0.6225	0.6211	0.6393	0.6295

3.2   Task 2: Prognostic Prediction Results

The individual performance of 10 conventional features in the training set by five folds cross validation is listed in Table 3. Their performances in testing set were not verified due to the limited number of submissions. After feature selection, two conventional features (number of nodes, TLG2.5) were retained to build prognostic model 1), which showed a C-index value of 0.658 in the testing set. The overall results of our submitted three models are provided in Table 4. For radiomics features with Combat harmonization, eight features (listed in Table 4) were combined in Model 2, obtaining a C-index value of 0.645 in the testing set. For radiomics features without harmonization, only two features were selected by the designed feature selection strategy. We further combined them with two conventional features to build prognostic model 3, which showed an improved performance in the training set but slightly decreased performance in testing (0.648).

Table 4. The performance of conventional, radiomics and combined models for RFS prediction.

Model	Features	Validation (5-fold)	Training	Testing
Conventional	number of nodes, TLG2.5	0.6716	0.6590	0.6582
Radiomics (with Combat)	CT-Morphology-spherical disproportion, CT-Local-peak, CT-IH-qcod, PET-Statistic-skewness, CT-GLSZM-szhge, PET-GLDZM-zdnu, PET-NGTDM-coarseness, PET-NGLDM-lgce	0.6453	0.6777	0.6452
Conventional+ Radiomics (without Combat)	number of nodes, TLG2.5, CT-Local-peak, PET-GLCM-correlation1	0.6810	0.6860	0.6475

Aggregation: IH, Intensity Histogram; qcod, quartile coefficient of dispersion; szhge, small zone high grey level emphasis; zdnu, zone distance non-uniformity; lgce, low grey level count emphasis.

4   Discussion

This study developed a fully automatic and relatively robust approach for segmentation and prognosis prediction based on primary tumor and lymph nodes of HNC patients using pre-treatment PET/CT imaging. Two segmentation models using 3D nnU-Net with and without pseudo labelling learning were trained. No improvement was found with PLL. Exploiting the predicted segmentation maks predicted by the nnU-Net, we extracted conventional and radiomics separately and developed three prognostic models via machine learning, to predict RFS. Our best submission obtained a C-index of 0.658.

In order to develop robust segmentation and prognosis models, we mainly focused on the relatively reliable and simple methods, such as the self-configurating nnU-Net and conventional features, which were validated by a vast of previous multi-center studies [13, 23]. Our segmentation task further proves the power of nnU-Net in medical images segmentation, although it did not perform at the top of the rankings in this challenge. [25] introduced the squeeze & excitation structure to nnU-Net and successfully improved their performance in last year's challenge. Moreover, attention mechanism was demonstrated with potential in segmentation when combining with convolutional networks [8, 11], which deserves further research. The use of pseudo labelling did not improve the performance in testing. We could define a threshold value to try identifying which test case can be used for re-training: For each prediction result, we can generate a confidence map, where the value of each pixel is the maximum probability. It can be considered a pseudo-label if no value in the whole confidence map is less than the threshold. This method needs to be further explored and optimized.

Our prognostic models based on whatever conventional and radiomics features showed relatively satisfying performance in the multi-center testing set. Our best model employed two conventional features which are number of nodes and TLG2.5 (MTV2.5 × SUVmean) with relatively high stability and generalization. The individual performance of each conventional features in the training set (Table 3) emphasized the importance of lymph nodes in prognosis prediction including its volume, number, and SUV metrics. This point is consistent with previous research [5, 15]. We did not perform harmonization for conventional features since they were demonstrated with high stability by previous analysis [18]. On the other hand, radiomics features after Combat harmonization also exhibit promising results with the C-index of 0.645 in the testing set. An interesting finding is that the testing result is very close to the mean performance of five folds cross validation (Table 4), which may potentially indicate the reduction of multi-center variations and the improvement of model generalization after Combat harmonization. Given the limitation of the number of submissions, we could not fully explore the effect of Combat harmonization through various strategies (harmonization by center, by clinical variables, separately for various features categories, or using unsupervised clustering for example). This will be further explored, as we have already done so with the HECKTOR 2021 set [20, 26]. Radiomics features without harmonization combined with two conventional features above did not improve the performance in testing, which may be explained by the high sensitivity of radiomics features to heterogeneity in the acquisition/reconstruction settings. Of note, our radiomics features were extracted from the regions including primary tumor and lymph nodes, rather than from primary tumor and lymph nodes separately, which will introduce different descriptions of tumor heterogeneity that need to be further explored.

5   Conclusion

In conclusion, our approach provides an automatic, fast and relatively consistent solution for primary tumor and lymph nodes segmentation in HNC patients, and shows potentials to be generally applied for prognosis evaluation by adopting both conventional and radiomics features.

References

1. Andrearczyk, V., et al.: Overview of the HECKTOR challenge at MICCAI 2021 automatic head and neck tumor segmentation and outcome prediction in PET/CT images. In: Andrearczyk, V., Oreiller, V., Hatt, M., Depeursinge, A. (eds.) 3D Head and Neck Tumor Segmentation in PET/CT Challenge, pp. 1–37. Springer, Cham (2021). https://doi.org/10.1007/978-3-030-98253-9_1
2. Andrearczyk, V., Oreiller, V., Depeursinge, A.: Oropharynx detection in PET-CT for tumor segmentation. Irish Mach. Vis. Image Process. **188** (2020)
3. Andrearczyk, V., et al.: Overview of the hecktor challenge at MICCAI 2022: automatic head and neck tumor segmentation and outcome prediction in pet/ct. In: Head and Neck Tumor Segmentation and Outcome Prediction. Springer, Cham (2023)
4. Bogowicz, M., et al.: Comparison of pet and CT radiomics for prediction of local tumor control in head and neck squamous cell carcinoma. Acta Oncol. **56**(11), 1531–1536 (2017)
5. Bogowicz, M., Tanadini-Lang, S., Guckenberger, M., Riesterer, O.: Combined CT radiomics of primary tumor and metastatic lymph nodes improves prediction of loco-regional control in head and neck cancer. Sci. Rep. **9**(1), 1–7 (2019)
6. Bonner, J.A., et al.: Radiotherapy plus cetuximab for locoregionally advanced head and neck cancer: 5-year survival data from a phase 3 randomised trial, and relation between cetuximab-induced rash and survival. Lancet Oncol. **11**(1), 21–28 (2010)
7. Chajon, E., et al.: Salivary gland-sparing other than parotid-sparing in definitive head-and-neck intensity-modulated radiotherapy does not seem to jeopardize local control. Radiat. Oncol. **8**(1), 1–9 (2013)
8. Chen, J., et al.: Transunet: transformers make strong encoders for medical image segmentation (2021). https://doi.org/10.48550/ARXIV.2102.04306, https://arxiv.org/abs/2102.04306
9. Goel, R., Moore, W., Sumer, B., Khan, S., Sher, D., Subramaniam, R.M.: Clinical practice in PET/CT for the management of head and neck squamous cell cancer. Am. J. Roentgenol. **209**(2), 289–303 (2017)
10. Harrell, F.E., Jr., Lee, K.L., Mark, D.B.: Multivariable prognostic models: issues in developing models, evaluating assumptions and adequacy, and measuring and reducing errors. Stat. Med. **15**(4), 361–387 (1996)
11. Hatamizadeh, A., et al.: Unetr: transformers for 3d medical image segmentation (2021). https://doi.org/10.48550/ARXIV.2103.10504, https://arxiv.org/abs/2103.10504
12. Isensee, F., Jaeger, P.F., Kohl, S.A., Petersen, J., Maier-Hein, K.H.: nnU-Net: a self-configuring method for deep learning-based biomedical image segmentation. Nat. Methods **18**(2), 203–211 (2021)
13. Isensee, F., et al.: nnU-Net: self-adapting framework for u-net-based medical image segmentation (2018). https://doi.org/10.48550/ARXIV.1809.10486, https://arxiv.org/abs/1809.10486

14. Johnson, W.E., Li, C., Rabinovic, A.: Adjusting batch effects in microarray expression data using empirical bayes methods. Biostatistics **8**(1), 118–127 (2007)
15. Kubicek, G.J., et al.: FDG-PET staging and importance of lymph node SUV in head and neck cancer. Head Neck Oncology **2**(1), 1–7 (2010)
16. Kumar, N., Verma, R., Sharma, S., Bhargava, S., Vahadane, A., Sethi, A.: A dataset and a technique for generalized nuclear segmentation for computational pathology. IEEE Trans. Med. Imaging **36**(7), 1550–1560 (2017)
17. Lee, D.H., et al.: Pseudo-label: the simple and efficient semi-supervised learning method for deep neural networks. In: Workshop on Challenges in Representation Learning, ICML, vol. 3, p. 896 (2013)
18. Leijenaar, R.T., et al.: Stability of FDG-PET radiomics features: an integrated analysis of test-retest and inter-observer variability. Acta Oncol. **52**(7), 1391–1397 (2013)
19. Murugesan, G.K., et al.: Head and neck primary tumor segmentation using deep neural networks and adaptive ensembling. In: Andrearczyk, V., Oreiller, V., Hatt, M., Depeursinge, A. (eds.) 3D Head and Neck Tumor Segmentation in PET/CT Challenge, vol. 13209, pp. 224–235. Springer, Cham (2021). https://doi.org/10.1007/978-3-030-98253-9_21
20. Abdallah, N., et al.: Predicting progression-free survival from FDG PET/CT images in head and neck cancer : comparison of different pipelines and harmonization strategies in the HECKTOR 2021 challenge dataset. In: 2022 IEEE Nuclear Science Symposium and Medical Imaging Conference (NSS/MIC). IEEE (2022)
21. Oreiller, V., et al.: Head and neck tumor segmentation in PET/CT: the HECKTOR challenge. Med. Image Anal. **77**, 102336 (2022)
22. Parkin, D.M., Bray, F., Ferlay, J., Pisani, P.: Global cancer statistics 2002. CA: Cancer J. Clin. **55**(2), 74–108 (2005)
23. Picchio, M., et al.: Predictive value of pre-therapy 18F-FDG PET/CT for the outcome of 18F-FDG pet-guided radiotherapy in patients with head and neck cancer. Eur. J. Nucl. Med. Mol. Imaging **41**(1), 21–31 (2014). https://doi.org/10.1007/s00259-013-2528-2
24. Vallieres, M., et al.: Radiomics strategies for risk assessment of tumour failure in head-and-neck cancer. Sci. Rep. **7**(1), 1–14 (2017)
25. Xie, J., Peng, Y.: The head and neck tumor segmentation using nnU-Net with Spatial and Channel 'Squeeze & Excitation' Blocks. In: Andrearczyk, V., Oreiller, V., Depeursinge, A. (eds.) HECKTOR 2020. LNCS, vol. 12603, pp. 28–36. Springer, Cham (2021). https://doi.org/10.1007/978-3-030-67194-5_3
26. Xu, H., Lu, L., Hatt, M.: Comparison of progressive combat for harmonization of radiomics features in multi-center head and neck tumor FDG PET/CT dataset from HECKTOR challenge 2021 (2022)
27. Zwanenburg, A., Leger, S., Vallières, M., Löck, S.: Image biomarker standardisation initiative. arxiv 2016. arXiv preprint arXiv:1612.07003

MLC at HECKTOR 2022: The Effect and Importance of Training Data When Analyzing Cases of Head and Neck Tumors Using Machine Learning

Vajira Thambawita[1]([⊠]), Andrea M. Storås[1,2], Steven A. Hicks[1]([⊠]),
Pål Halvorsen[1,2], and Michael A. Riegler[1,3]

[1] SimulaMet, Oslo, Norway
{vajira,steven}@simula.no
[2] Oslo Metropolitan University, Oslo, Norway
[3] UiT The Arctic University of Norway, Tromsø, Norway

Abstract. Head and neck cancers are the fifth most common cancer worldwide, and recently, analysis of Positron Emission Tomography (PET) and Computed Tomography (CT) images has been proposed to identify patients with a prognosis. Even though the results look promising, more research is needed to further validate and improve the results. This paper presents the work done by team MLC for the 2022 version of the HECKTOR grand challenge held at MICCAI 2022. For Task 1, the automatic segmentation task, our approach was, in contrast to earlier solutions using 3D segmentation, to keep it as simple as possible using a 2D model, analyzing every slice as a standalone image. In addition, we were interested in understanding how different modalities influence the results. We proposed two approaches; one using only the CT scans to make predictions and another using a combination of the CT and PET scans. For Task 2, the prediction of recurrence-free survival, we first proposed two approaches, one where we only use patient data and one where we combined the patient data with segmentations from the image model. For the prediction of the first two approaches, we used Random Forest. In our third approach, we combined patient data and image data using XGBoost. Low kidney function might worsen cancer prognosis. In this approach, we therefore estimated the kidney function of the patients and included it as a feature. Overall, we conclude that our simple methods were not able to compete with the highest-ranking submissions, but we still obtained reasonably good scores. We also got interesting insights into how the combination of different modalities can influence the segmentation and predictions.

1 Introduction

Head and neck cancers are among the most common cancer types worldwide. Early detection is critical as the tumor's size on diagnosis will dictate the patient's quality of life and chances of survival [10]. Medical image analysis

V. Andrearczyk et al. (Eds.): HECKTOR 2022, LNCS 13626, pp. 166–177, 2023.
https://doi.org/10.1007/978-3-031-27420-6_17

and radiomics have shown promising results in detecting different diseases and cancers [2,9,16], including those found in the head and neck [18,20,25]. In this paper, we describe our approaches for the HEad and neCK TumOR (HECK-TOR) grand challenge held at MICCAI 2022 [1,17]. Of the two tasks presented at the challenge, we participated in both. In Task 1, the aim was to segment tumors from Computed Tomography (CT) and Positron Emission Tomography (PET) scans of the head and neck (examples shown in Fig. 1). Task 2 asked for the prediction of Recurrence-Free Survival (RFS) based on clinical information about the patients presented in a tabular format, which also could be combined with the outputs from Task 1.

As the provided dataset contained different types of data, our strategy to tackle the HECKTOR challenge was to explore how the inclusion and combination of different modalities change the prediction outcome. In this respect, we investigated how CT and PET scans can be used individually or combined for tumor segmentation in Task 1 and how RFS can be predicted using the metadata with or without tumor information for Task 2. The main contributions of this paper are as follows:

1. A comparison of simple segmentation methods using CT or PET slices individually or combined.
2. Understanding the effect of combining different data modalities on the analysis results.
3. Analysis of what features were most relevant for predicting RFS using patient-related data and image features.

2   Methods

In this section, we describe the methods we applied to solve Tasks 1 and 2, respectively.

2.1   Task 1: Segmentation of CT and PET Scans

For Task 1, we used the provided development dataset consisting of CT scans, PET scans, and corresponding segmentation masks. We experimented with three different approaches as follows (see Fig. 2):

Approach 1: Only using individual slices of the CT scans to predict tumors with a UNet++-based model [28].

Approach 2: Combining the CT and PET scans by stacking CT, PET, and the mean of CT and PET images channel-wise and passing them through a UNet++-based model.

Approach 3: Analyzing CT and PET slices separately in an ensemble-like setup using a TriUnet-based model [24].

These three approaches utilize the data provided in the HECKTOR competition differently, from simple to more complex. The following sections describe all the steps of data pre-processing, sampling, augmentation, implementation of the models, and post-processing.

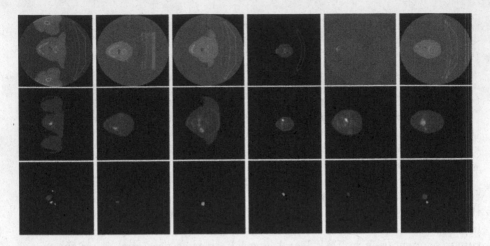

Fig. 1. Six example data points from the development dataset provided by the HECK-TOR organizers. The first row contains a slice from one of the CT images, the second row contains a slice from one of the PET images, and the third row contains the corresponding segmentation mask. Please note that the brightness of the PET images has been adjusted so that the contents are more easily visible.

Image Data Pre-processing: We divided the development dataset into a training and a validation dataset containing 90% and 10% samples, respectively. For Approach 1, we extracted the slices from the CT and ground truth as *.png* images without applying re-sampling because the shape of the CT and provided ground truth were the same. However, for Approaches 2 and 3, we performed slice extraction after re-sampling (using SimpleITK [26]). We used the same re-sampling as provided by the task organizers[1] with default spacing $(2, 2, 2)$. In addition, we normalized all CT and PET images into the range between 0 and 255, but not ground truth images that contain pixel values of $[0, 1, 2]$. After the extraction process, we noticed that the training dataset contained large number of true negative samples. Therefore, to avoid bias, we re-balanced the training dataset by extracting only slices with true positive pixels for H&N Primary tumors (GTVp) and H&N nodal Gross Tumor (GTVn). The class rebalancing was done by combining an equal number of true positive slices with the true negative slices extracted from the initial training dataset. To make a challenging validation dataset, we extracted only slices with GTVp and GTVn from the validation images.

We applied similar image augmentation for all three approaches. The Albumentations [6] library provides a set of augmentation options for image segmentation tasks. More information about the input parameters of the augmentation methods can be found in our GitHub repository[2].

---

[1] https://github.com/voreille/hecktor/blob/master/src/resampling/resample_2022.py.
[2] https://github.com/vlbthambawita/hecktor_2022_MLC.

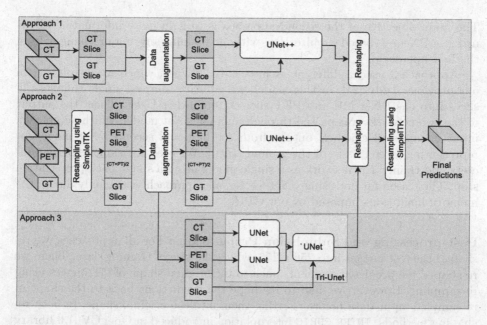

Fig. 2. Three approaches used for Task 1. Approach 1: uses only CT images and the corresponding ground truth (GT). Approach 2: inputs stack of CT, PET, mean of CT and PET. Approach 3: uses two separate UNet models for CT and PET and another UNet for final predictions. Reshaping sizes used in Approach 1 is different from the sizes used for Approach 2 and 3.

Model Architectures, Hyperparameters and Inputs: The models for Task 1 were implemented in Pytorch [19] using the Segmentation Models library [12]. All models were trained for 100 epochs on hardware consisting of two Nvidia RTX 3080 Graphic Processing Units (GPUs) with 10 GB of memory each, an AMD Ryzen 9 3950X 16-core processor, and 64 GB memory. Submissions were made with the best-performing checkpoints, which were selected based on the performance on the validation dataset. For the first 50 epochs, the learning rate was set to 0.0001, then reduced to 0.00001 for the remaining 50 epochs. The Adam optimizer [14] with default parameters except the learning rate was used for all the experiments. Furthermore, we have used DiceLoss with skipped channel 0 as the main loss function in the training process and the Intersection over Union (IoU) as a metric to evaluate our models.

In Approach 1, we have used a UNet++-based model with $se\_resnext50\_32 \times 4d$ as the encoder. The model was trained using only single channel CT input images and the corresponding ground truth masks after resizing them into 256×256 in the augmentation step.

For Approach 2, we re-sampled the CT and PET slices and trained a UNet++ model. For this approach, we stack a CT slice, a PET slice, and the mean of the CT and PET slice in the color channel and use these as input to the model.

The main objective of the second approach is to gain more information about using a UNet++-based architecture without making any major changes from the first approach.

Approach 3 used a different architecture, TriUnet, which we introduced in our previous study [24]. In this model, we input re-sampled single channel CT slices into one UNet [21] and PET slices into another UNet. Then, the output of the two networks was passed through another UNet model, which accepts six input channels (3 channels output from the first and second UNet model for representing three classes of the ground truth). We used the same hyperparameters and trained the network as a single model using a single back-propagation step. The reason for not using UNet++ for this approach was mainly due to the memory limitations imposed by our GPU.

Post-processing and Submission Preparations: For all approaches, we re-shaped the test images into 256×256, which the size of training data. Then, we re-shaped the predicted segments back to the original shape of CT images using re-sampling. However, we had to re-shape the predictions back to the shape of re-sampled input data before re-sizing them into the original shape. In both re-shaping methods, INTER_CUBIC interpolation introduced in OpenCV [13] library was used.

For all approaches in Task 1, we used the academic version of Weights and Biases [3] for tracking and analyzing experiments and the corresponding performance. All the experiments with the corresponding best checkpoints are available on GitHub[3].

2.2   Task 2: Prediction of Recurrence-Free Survival

Estimation of Kidney Function. We include the estimated kidney function as a feature for the XGBoost model from the third approach of Task 2 as this might improve the predictions of RFS. Prior research indicates that chronic kidney disease can worsen the prognosis of cancer patients and that monitoring the kidney function of cancer patients is crucial [22].

The feature is created using the Cockraft-Gault formula, which is among the most widely used formulas for estimating the kidney function [8]. This formula requires the gender, age, body weight and serum creatinine concentration. Because serum creatinine is not available in the dataset, the average values for men and women are used instead [23]. Indeed, when plotting the correlation matrix for the training data, we observe that there is a positive correlation between the estimated kidney function and the RFS (correlation = 0.26), indicating that higher kidney function is associated with a longer time to recurrence. The entire correlation matrix for the training dataset is shown in Fig. 5.

---

[3] https://github.com/vlbthambawita/hecktor_2022_MLC.

Description of the Approaches. For Task 2, we proposed three different approaches. The first approach used only the patient data, while the second and third approach also included features based on the image data. The image features arrived from the segmentation masks from the best approach of Task 1. Specifically, we calculated the number of pixels per class from the predicted masks in addition to the number of slices of the CT images in the z-dimension, resulting in four additional features. Moreover, the third approach used the estimated kidney function as a feature. For all three approaches, we used 10-fold cross-validation on the development data to determine the best hyperparameters and model. The hyperparameters are selected based on the root mean squared error (RMSE) of the model, which should be as low as possible. The final models are trained on the entire training dataset using the identified set of hyperparameters. The resulting models are then used on the test dataset to make the predictions for the challenge evaluation. All experiments are performed using the scikit-learn library [5].

Approach 1: The first approach used the Random Forest [4] algorithm to predict RFS using only the patient data. Random Forest was chosen because it is known to work well on tabular data and is often used as a baseline for medical-related machine learning problems. All features provided in the training data were used besides the patient ID. Based on the cross-validation results (RMSE of 988.47), the hyperparameters for the Random Forest were set as the following; max features as the number of features divided by three, and the number of trees was set to 100. All other hyperparameters used the default value set by scikit-learn.

Approach 2: For the second approach, we used the same algorithm as the first approach, but with additional image features as described in the beginning of the subsection. The RMSE from the cross-validation of the training data was 962.83, which was an improvement compared to the first approach showing that the inclusion of image data has a positive effect on the results. The hyperparameters used for the Random Forest in the second approach are as follows; max features as the number of features and the number of trees 200. All other hyperparameters used the default value as set by scikit-learn.

Approach 3: Regarding the third approach, an XGBoost [7] regression model was trained to predict RFS using the available patient data and the image features with one additional feature representing the estimated kidney function. The feature representing alcohol consumption was removed because the majority of the patients in both training and test dataset do not have any registered value for this feature. The patient ID was not included in the training dataset. The RMSE from the cross-validation on the training data was 909.09. The hyperparameters for the XGBoost model are: 'n_estimators' = 120, 'learning_rate' = 0.05, 'max_depth' = 4, 'subsample' = 0.7, 'colsample_bytree' = 0.6, 'colsample_bynode' = 1 and 'colsample_bylevel' = 0.8. The other hyperparameters used the default value.

Investigating Feature Importance. After training the XGBoost model, feature importance is explored using Shapley additive explanations (SHAP) [15].

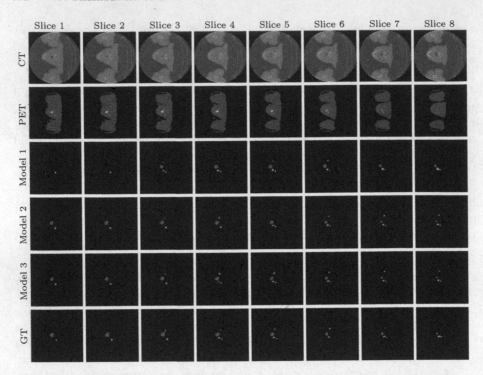

Fig. 3. Example predictions taken from each model at different slices and the corresponding ground truth.

Table 1. Official HECKTOR Challenge 2022 results for Task 1 using the mean aggregated Dice metric for the ranking. The best values are marked using **bold text**.

ID	Model	Dataset	Validation			Test		
			GTVp	GTVn	Mean	GTVp	GTVn	Mean
1	UNet++	CT	0.560	0.721	0.641	0.466	0.536	0.501
2	UNet++	CT + PET	0.601	0.674	0.638	0.607	0.604	0.605
3	TriUnet	CT + PET	**0.671**	**0.722**	**0.696**	**0.659**	**0.654**	**0.657**

SHAP approximates Shapley values, which origin from game theory and assigns values to the features based on how much they contribute to the prediction [15,27]. Consequently, it is possible to investigate which features the model regards as most important.

3   Discussion and Results

3.1   Task 1

Looking at the results for Task 1 in Table 1, we see that the first approach struggles to detect GTVp in both the validation and the test datasets. This is

Table 2. Official HECKTOR Challenge 2022 results for Task 2 using C-index for the ranking. The best values are marked using **bold text**.

ID	Model	Data	C-index
1	Random Forest	Patient data	0.585
2	Random Forest	Patient and image data	0.589
3	XGBoost	Patient and image data	**0.656**

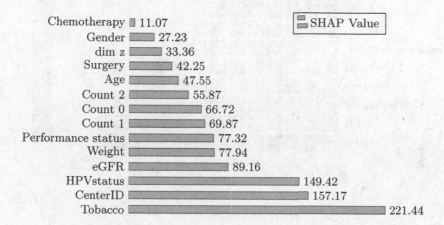

Fig. 4. Feature importance for the XGBoost model from Task 2, trained on image features and patient information. Count 0, Count 1, Count 2 and dim z are image features.

also shown in Fig. 3, where the first model is unable to detect the presence of GTVp until the third slice. Despite not being able to detect GTVp very well, the first approach performs well on segmenting GTVn on the validation dataset, but not on the test dataset. This can most likely be attributed to the differences between the validation and test datasets, as the other approaches show similar results. Adding information from the PET scans for the latter two approaches seems to help in detecting GTVp, as evident by the improved scores in Table 1, and they are both able to detect GTVp in all eight slices from the example in Fig. 3. The differences between Approaches 2 and 3 indicate that extracting features from the CT and PET scans independently seems to be the most suitable technique.

3.2   Task 2

For Task 2, the results can be seen in Table 2. From the results, we can observe several interesting insights. Firstly, adding additional data to the patient data gives better predictions. This can be observed in the difference between Approaches 1 and 2 when image data was added as additional features. We can

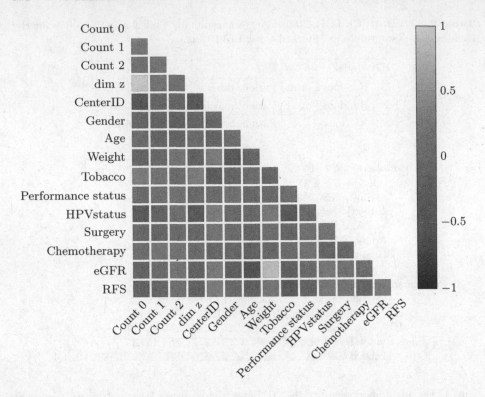

Fig. 5. Correlation matrix for the training data applied in Task 2 for the third approach. Count 0, Count 1, Count 2, and dim z are image features.

also observe that XGBoost outperforms Random Forest by a large margin. This correlates with general findings in the literature that XGBoost is one of the best working methods. This questions also the general concept of using Random Forest as a baseline and suggest that in general it should be replaced with XGBoost instead.

The SHAP values estimated for the third approach are plotted in Fig. 4 and give us a better understanding of which features are most relevant to the model. We observe that the top five features are Tobacco, CenterID, HPVstatus, estimated glomerular filtration rate (eGFR) which represent the kidney function, and Weight (most important first). Tobacco consumption is the most important feature. This is not surprising as it is well-known that tobacco increases the risk of developing head and neck cancer, see for example [11]. The kidney function is ranked as number four and seems to be an important indicator for RFS. This finding is in line with earlier research, where a relationship between kidney function and prognoses for cancer patients has been identified [22]. An important limitation is that the true serum creatinine values were not available in the provided dataset. The creatinine concentration will to a large extent affect the estimated kidney function, and only applying the average gender values is not enough for getting accurate individual estimations. Despite this, we believe that serum creatinine should be included in future datasets to see if the model performance can be further improved. Interestingly, CenterID is ranked as the second most important feature, which is also confirmed by a positive correlation between the CenterID and RFS in the correlation plot (Fig. 5). These observations might be due to different patient populations at different centers, e.g., there might be more severely ill patients treated at one center while less serious cases are treated at another center. Another possibility is that the medical doctors choose different strategies for treating the patients or that the surgical skills differ between the centers. However, neither Surgery or Chemotherapy are among the highest-ranked features. The CenterID was not encoded as a categorical feature. If this had been done, the results from the SHAP analysis might change. The five least important features are Chemotherapy, Gender, dim z, surgery, and Age (least important first). The image features rank in the middle range regarding feature importance showing that they can be an important factor in predicting RFS. Considering that our imaging method and the image features are simple, we assume that with more advanced image analysis methods and the corresponding resulting features, the importance of the image-related features might increase even further.

4 Conclusion

In conclusion, our simple methods were not able to perform at the same levels as the highest ranked submissions despite achieving reasonable results. We were able to obtain some interesting insights regarding combinations of different data modalities, showing that the combination of different sources improves the results even when simple methods are used. For Task 2, we also had a closer look at feature importance, revealing some interesting features such as the usefulness of eGFR. For future work, it would be interesting to apply the feature importance analysis to other solutions of the competition to investigate if they are leading to similar findings. Furthermore, we would also like to investigate the CenterID correlation to explore if a hospital-specific treatment or country-related factor is influencing it.

References

1. Andrearczyk, V., et al.: Overview of the HECKTOR challenge at MICCAI 2022: automatic head and neck tumor segmentation in PET/CT. In: Head and Neck Tumor Segmentation and Outcome Prediction (2022)
2. Bandyk, M.G., Gopireddy, D.R., Lall, C., Balaji, K., Dolz, J.: MRI and CT bladder segmentation from classical to deep learning based approaches: Current limitations and lessons. Comput. Biol. Med. **134**, 104472 (2021)
3. Biewald, L.: Experiment tracking with weights and biases (2020). https://www.wandb.com/
4. Breiman, L.: Random forests. Mach. Learn. **45**(1), 5–32 (2001)

5. Buitinck, L., et al.: API design for machine learning software: experiences from the scikit-learn project. In: ECML PKDD Workshop: Languages for Data Mining and Machine Learning, pp. 108–122 (2013)

6. Buslaev, A., Iglovikov, V.I., Khvedchenya, E., Parinov, A., Druzhinin, M., Kalinin, A.A.: Albumentations: fast and flexible image augmentations. Information **11**(2) (2020). https://doi.org/10.3390/info11020125, https://www.mdpi.com/2078-2489/11/2/125

7. Chen, T., Guestrin, C.: XGBoost: a scalable tree boosting system. In: Proceedings of the ACM SIGKDD International Conference on Knowledge Discovery and Data Mining (KDD), pp. 785–794 (2016). https://doi.org/10.1145/2939672.2939785

8. Cockcroft, D.W., Gault, H.: Prediction of creatinine clearance from serum creatinine. Nephron **16**, 31–41 (1976). https://doi.org/10.1159/000180580

9. Duran, A., Dussert, G., Rouvière, O., Jaouen, T., Jodoin, P.M., Lartizien, C.: Prostattention-net: a deep attention model for prostate cancer segmentation by aggressiveness in MRI scans. Med. Image Anal. **77**, 102347 (2022)

10. Gerstner, A.: Early detection in head and neck cancer - current state and future perspectives. GMS Current Topics Otorhinolaryngol. Head Neck Surg. **7** (2008)

11. Hashibe, M., et al.: Interaction between tobacco and alcohol use and the risk of head and neck cancer: pooled analysis in the international head and neck cancer epidemiology consortium. Cancer Epidemiol. Biomarkers Prev. **18**(2), 541–550 (2009). https://doi.org/10.1158/1055-9965.EPI-08-0347

12. Iakubovskii, P.: Segmentation models pytorch. https://github.com/qubvel/segmentation_models.pytorch (2019)

13. Itseez: Open source computer vision library. https://github.com/itseez/opencv (2015)

14. Kingma, D.P., Ba, J.: Adam: a method for stochastic optimization. arXiv preprint arXiv:1412.6980 (2014)

15. Lundberg, S.M., et al.: Explainable AI for trees: from local explanations to global understanding (2019). https://doi.org/10.48550/arXiv.1905.04610

16. Mittmann, B.J., et al.: Deep learning-based classification of DSA image sequences of patients with acute ischemic stroke. Int. J. Comput. Assist. Radiol. Surg. **17**, 1–9 (2022). https://doi.org/10.1007/s11548-022-02654-8

17. Oreiller, V., et al.: Head and neck tumor segmentation in PET/CT: the HECKTOR challenge. Med. Image Anal. **77**, 102336 (2022). https://doi.org/10.1016/j.media.2021.102336

18. Outeiral, R.R., et al.: Strategies for tackling the class imbalance problem of oropharyngeal primary tumor segmentation on magnetic resonance images. Phys. Imaging Radiat. Oncol. **23**, 144–149 (2022)

19. Paszke, A., et al.: Pytorch: an imperative style, high-performance deep learning library. In: Wallach, H., Larochelle, H., Beygelzimer, A., d'Alché-Buc, F., Fox, E., Garnett, R. (eds.) Advances in Neural Information Processing Systems 32, pp. 8024–8035. Curran Associates, Inc. (2019). http://papers.neurips.cc/paper/9015-pytorch-an-imperative-style-high-performance-deep-learning-library.pdf

20. Ren, J., Eriksen, J.G., Nijkamp, J., Korreman, S.S.: Comparing different CT, pet and MRI multi-modality image combinations for deep learning-based head and neck tumor segmentation. Acta Oncol. **60**(11), 1399–1406 (2021)

21. Ronneberger, O., Fischer, P., Brox, T.: U-Net: convolutional networks for biomedical image segmentation. In: Navab, N., Hornegger, J., Wells, W.M., Frangi, A.F. (eds.) MICCAI 2015. LNCS, vol. 9351, pp. 234–241. Springer, Cham (2015). https://doi.org/10.1007/978-3-319-24574-4_28

22. e Silva, V.T.D.C., Costalonga, E.C., Coelho, F.O., Caires, R.A., Burdmann, E.A.: Assessment of kidney function in patients with cancer. Adv. Chronic Kidney Dis. **25**(1), 49–56 (2018)

23. Staff, M.C.: Creatinine tests. https://www.mayoclinic.org/tests-procedures/creatinine-test/about/pac-20384646 (2021)

24. Thambawita, V.L.B., Hicks, S., Halvorsen, P., Riegler, M.: Divergentnets: Medical image segmentation by network ensemble. In: Proceedings of the International Workshop and Challenge on Computer Vision in Endoscopy (EndoCV), pp. 27–38 (2021)

25. Wahid, K.A.: Evaluation of deep learning-based multiparametric MRI oropharyngeal primary tumor auto-segmentation and investigation of input channel effects: Results from a prospective imaging registry. Clin. Transl. Radiat. Oncol. **32**, 6–14 (2022)

26. Yaniv, Z., Lowekamp, B.C., Johnson, H.J., Beare, R.: Simpleitk image-analysis notebooks: a collaborative environment for education and reproducible research. J. Digit. Imaging **31**(3), 290–303 (2018)

27. Young, H.P.: Monotonic solutions of cooperative games. Internat. J. Game Theory **14**, 65–72 (1985). https://doi.org/10.1007/BF01769885

28. Zhou, Z., Siddiquee, M.M.R., Tajbakhsh, N., Liang, J.: Unet++: redesigning skip connections to exploit multiscale features in image segmentation. IEEE Trans. Med. Imaging **39**(6), 1856–1867 (2019)

Towards Tumour Graph Learning for Survival Prediction in Head & Neck Cancer Patients

Ángel Víctor Juanco-Müller[1,2]([✉]) [ID], João F. C. Mota[2] [ID], Keith Goatman[1] [ID], and Corné Hoogendoorn[1] [ID]

[1] Canon Medical Research Europe Ltd., Edinburgh, UK
victor.juancomuller@mre.medical.canon
[2] Heriot-Watt University, Edinburgh, UK

Abstract. With nearly one million new cases diagnosed worldwide in 2020, head & neck cancer is a deadly and common malignity. There are challenges to decision making and treatment of such cancer, due to lesions in multiple locations and outcome variability between patients. Therefore, automated segmentation and prognosis estimation approaches can help ensure each patient gets the most effective treatment. This paper presents a framework to perform these functions on arbitrary field of view (FoV) PET and CT registered scans, thus approaching tasks 1 and 2 of the HECKTOR 2022 challenge as team VokCow. The method consists of three stages: localization, segmentation and survival prediction. First, the scans with arbitrary FoV are cropped to the head and neck region and a u-shaped convolutional neural network (CNN) is trained to segment the region of interest. Then, using the obtained regions, another CNN is combined with a support vector machine classifier to obtain the semantic segmentation of the tumours, which results in an aggregated Dice score of 0.57 in task 1. Finally, survival prediction is approached with an ensemble of Weibull accelerated failure times model and deep learning methods. In addition to patient health record data, we explore whether processing graphs of image patches centred at the tumours via graph convolutions can improve the prognostic predictions. A concordance index of 0.64 was achieved in the test set, ranking 6th in the challenge leaderboard for this task.

1 Introduction

Tumours occurring in the oropharyngeal region are commonly referred to as head and neck (H&N) Cancer. In 2020 they were the third most commonly diagnosed cancers worldwide [1]. To inform the difficult decisions that oncologists often have to make, prognosis estimation has been shown to result in better treatment planning and improved patient quality of life [2]. Therefore, automatic lesion segmentation and risk score prediction algorithms have the potential to speed up clinicians workloads, enabling them to treat more patients.

V. Andrearczyk et al. (Eds.): HECKTOR 2022, LNCS 13626, pp. 178–191, 2023.
https://doi.org/10.1007/978-3-031-27420-6_18

The HECKTOR challenge was conceived [3,4] to advance the task of automatic primary tumour (GTVp) segmentation and prognosis prediction. Since the first edition in 2020, the dataset has increased from 254 to 325 cases in 2021 [5], and up to a total of 883 cases in the 2022 edition [6]. Other characteristics of the present release are the lack of region of interest (RoIs) and the inclusion of secondary lymph nodes (GTVn) as segmentation targets.

This paper describes a framework for tumour segmentation and prognosis prediction consisting of three stages. First, a localization model finds the neck region in the input scans (Sect. 2.1). Then, to obtain segmentation masks for task 1, we train a u-shaped convolutional neural network (UNet) [7,8] to distinguish between tumour and background, and a support vector machine (SVM) to predict the tumour type and discard false positives (Sect. 2.2), resulting in the semantic segmentation output for task 1, which achieves an average Dice score (DSC) of 0.57 in the test set. Finally, we explore combinations of the deep multi task logistic regression (MTLR) model, featuring CNNs and graph convolutions networks, and the Weibull accelerated failure times (Weibull AFT) method to predict the prognosis metric, which is the *relapse free survival* (Sect. 2.3), resulting in a concordance index of 0.64 in the test set.

The experimental implementation is detailed in Sect. 3, results are presented in Sect. 4, and a discussion of our findings is provided in Sect. 5.

2 Materials and Methods

This section presents the three main stages of our framework: localization Sect. 2.1, segmentation Sect. 2.2 and survival prediction Sect. 2.3, depicted in Fig. 1.

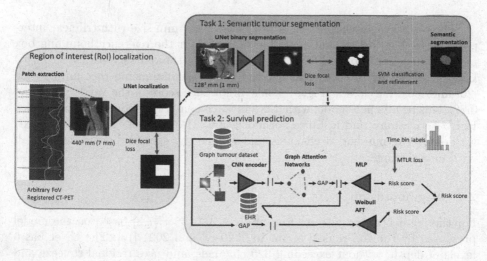

Fig. 1. The three main stages of the proposed framework. GAP stands for global average pooling, MLP for multi layer perceptron, and || refers to concatenation.

2.1   Localization

First, we extract 440^3 mm patches of the head and neck region from the arbitrary FoV PET-CT scan by analysing the CT and PET mean slice intensity along the z-axis. The brain is detected by a peak in the PET signal and the neck by an abrupt drop of the CT value. To avoid false positives caused by peaks of the PET signal in other regions of the body (e.g., bladder) we restrict the landmark search only to the first 250 mm starting from the head, as depicted in Fig. 2.

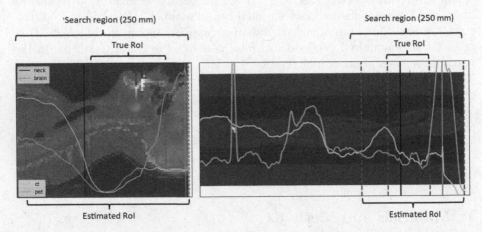

Fig. 2. Patch extraction for two different FoV cases. The dotted red lines are the inferred bounds, whereas the green ones correspond to the ground truth location of the RoI. (Color figure online)

The resulting patches are then resized to a 64^3 mm size with trilinear interpolation for the images and nearest neighbours for the reference bounding box masks, which are obtained from the ground truth tumour segmentations. A 3D UNet [7,8] is then used to segment the latter by minimizing the sum of Dice [20] and focal [22] losses (Fig. 3), achieving a Dice score in the validation set of 0.72. This results in a model with 3 layers, each comprising convolutional blocks, ReLU activations, and instance normalization [21]. From one layer to the next, we double the number of channels and reduce the spatial dimensions by half with max pooling.

2.2   Segmentation

For the segmentation task, we first apply a 3D UNet [7,8] based on the model presented by the top ranked teams in 2020 [9] and 2021 [10]. The UNet has 5 levels of depth, without exceeding 320 channels, and uses residual squeeze and excitation blocks [8]. The loss function was the same one used for localization.

Because a multi-class segmentation model performed poorly, we opted instead to use the UNet for binary segmentation (tumour-background), and a traditional

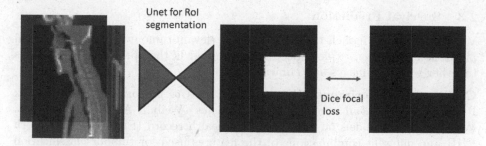

Fig. 3. RoI segmentation with UNet. Based on the bounds inferred in the previous step, the images are cropped to a common FoV and a UNet is trained on a low-resolution dataset to segment the target RoI bounding box.

classifier to infer the tumour type. We tried different algorithms and a support vector machine (SVM) with radial basis functions was chosen as it yielded the best performance. The input features were: *tumour centroid* and *bounding box coordinates, Euler number, extent, solidity, filled area, area of the convex hull, area of the bounding box, maximum Feret diameter, equivalent diameter area, eigenvalues of moment of inertia* and *minimum, mean* and *max values of CT and PET intensities.*

The SVM classifies the tumours into three possible classes: background (to discard false positive predictions), primary tumour (GTVp) and lymph node tumours (GTVn). The input features were extracted with the Scikit-Image library [12], and the SVM implementation is provided by the Scikit-Learn [13] library with default parameters. The overall pipeline is depicted in Fig. 4.

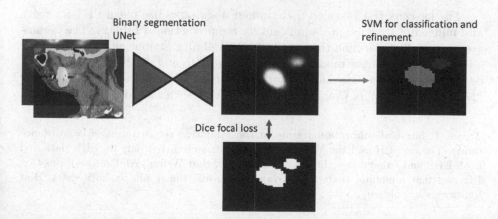

Fig. 4. Semantic tumour segmentation. First a UNet segments the tumours from the background and then an SVM classifier predicts the tumour type.

2.3   Survival Prediction

We first use simple models to select the most relevant tumour features for survival prediction, and then explore deep learning models that process such features together with the images and clinical data.

Calibration Experiments. We considered the *Cox proportional hazards* (Cox PH) [15] and *Weibull accelerated failure times* (Weibull AFT) [14] methods. We fitted these models on only electronic health record (EHR) data and both EHR and different combinations of the features used for tumour classification with SVM. For cases with several tumours, we considered the mean value of the features. We also included the number of tumours as an additional feature. After experimentation we identified *tumour centroid, mean CT and PET, max CT intensities*, and *number of tumours* as the combination yielding the best improvement of survival prediction in the validation set.

Among the EHR variables, there was missing information for some patients regarding alcohol and tobacco usage, performance status in the Zubrod scale, presence of human papillomavirus (HPV) and whether the patient has undergone surgery or not. Because all these variables are non-negative, we assigned the value -1 for the missing cases. This resulted in better performance than simply dropping these columns.

The Weibull AFT model trained on EHR data and tumour descriptors achieved the highest concordance index (C-Index) in the validation set (Table 1). This model assumes the hazard probability to be a function of patient features x_i and time t parametrized by β_i and ρ,

$$H(x_i, t; \beta_i, \rho) \propto - \left(\frac{t}{\exp(\beta_0 + \sum_i \beta_i x_i)} \right)^{\rho}.$$

The log-rank test revealed that tumour descriptors like mean PET intensity and number of tumours are significant for the predictions (Table 2). The parameters ρ and β_0 determine the shape of the Weibull distribution and were fit to -3.07 and 34.24 for the model trained only on EHR, and -0.21 and 10.25 for the one trained on both EHR and tumour descriptors. We used the Lifelines package [24] implementation of Cox PH and Weibull AFT with their default parameters.

Table 1. Survival calibration results in the validation set in terms of concordance index. The Cox PH and the Weibull AFT models were fitted only on EHR data and both EHR and tumour descriptors. It can be seen that Weibull AFT outperforms Cox PH and that including tumour descriptors improves the results in both cases. Best figures are in bold.

	Cox PH	Weibull AFT
EHR	0.58589	0.60996
EHR + tumour descriptors	0.60977	**0.63014**

Table 2. Log-rank test for the Weibull AFT model fit. The input features with lowest p-value are the most representative for the survival prediction task. Best figures in bold.

Input feature (x_i)	Only EHR		EHR + tumour descriptors	
	Coefficient (β_i)	p-value ($\downarrow$)	Coefficient (β_i)	p-value ($\downarrow$)
Age	−0.15	0.35	−0.13	0.23
Alcohol	0.36	0.10	0.44	0.07
Chemotherapy	−0.14	0.47	−0.15	0.82
Gender	−0.50	0.01	−0.28	0.06
HPV status ($0 = -, 1 = +$)	0.37	0.08	0.31	0.12
Performance status	−0.87	**≪0.005**	−0.69	**≪0.005**
Surgery	0.16	0.37	0.15	0.65
Tobacco	0.13	0.54	0.03	0.40
Weight	0.42	0.03	0.55	0.01
Area bounding box	–	–	−0.28	**≪0.005**
Centroid x coordinate	–	–	−0.64	0.22
Centroid y coordinate	–	–	−1.41	0.99
Centroid z coordinate	–	–	0.92	0.23
Max CT intensity	–	–	−0.27	0.05
Mean CT intensity	–	–	0.54	0.02
Mean PET intensity	–	–	−0.55	**≪0.005**
Number of tumours	–	–	−0.43	**≪0.005**

Survival Model. The winning method of the previous edition of the challenge, named *Deep Fusion* [25], consisted of a CNN encoder that takes a fused PET-CT image as an input, and outputs a feature embedding that is concatenated with patient EHR data. The final layer is a multi layer perceptron (MLP) connected with the multi task logistic regression (MTLR) loss, which can model individual risk scores accurately [16–18]. It divides the target time into bins for which survival scores are predicted, imposing constraints to deal with uncensored and censored events.

Since *number of tumours*, n, was one of the most representative features in the calibration experiments, we hypothesize that, rather than one single image patch, processing n fused PET-CT patches with graph convolution networks may provide stronger prognosis signals. In the proposed *Multi-patch* model, 64-dimensional patch embeddings were first obtained with a CNN layer followed by batch normalization [19], ReLU activation and average pooling. Next, we built an unweighted fully connected graph of n nodes, one for each of the image patches.

We assigned concatenations of the CNN embeddings of the image patches and the tumour descriptors selected in the *calibration experiments* as node features, and applied two layers of graph convolution and ReLU activation to reason over the tumour graph. The improved graph attention network (GATv2) [11] was chosen for its availability to perform dynamic node attention.

Finally, an average pooling layer generates global graph vector embeddings, which are then concatenated with each patient's EHR data. A Multi Layer Perceptron (MLP) produces the output logits, which are then used to compute the MTLR loss and the patient's risk score. The full pipeline is shown in Fig. 5.

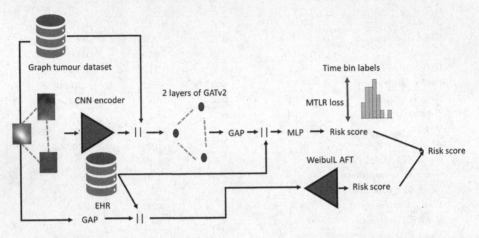

Fig. 5. Multi-patch network for survival prediction. The inputs are fused CT-PET image patches cropped at the segmented tumour centroids, tumour features, and patient EHR data. The output is the predicted risk score.

3   Experimental Set up

Here we provide details of our experimental implementation, covering data splitting (Sect. 3.1), the preprocessing and augmentation techniques (Sect. 3.2), and the hardware and network hyperparameters (Sect. 3.3).

3.1   Data Splitting

For the segmentation task, the training data was divided into training and validation splits with 445 and 79 cases respectively for the segmentation task. To ensure a balanced representation of multi-centre data, the cases from each centre were first randomly allocated to *per-centre* subsplits, and then aggregated to form the final split.

Since the survival dataset is a subset of the segmentation one, we opted to define the survival splits as subsets of the segmentation partitions, resulting in 414 training and 74 validation cases. In this manner the inferred risk score is determined by the segmentation inference. We confirmed (Fig. 6) the absence of important distribution shifts between the training and validation survivals times and proportion of censoring cases.

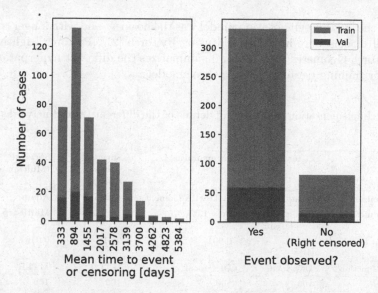

Fig. 6. Histograms of survival times and censoring status of the training and validation cases. It can be seen that both distributions are very similar, peaking at the same mean. Best seen in colour.

3.2 Data Preprocessing and Augmentations

Because global context is more important than detail for localization, we resampled the images to low resolution, e.g. $(7, 7, 7)$ mm voxel spacing and clipped the CT values to the interval $(-1024, 1024)$. Instead, for the segmentation and survival networks, where detail or texture is more important, we resampled the inputs to 1 mm isotropic voxel size and windowed the CT values to $(-200, 200)$ to enhance soft tissues.

In all cases, the images were normalized by subtracting the mean and dividing by the standard deviation. The input of the localization and segmentation networks are pairs of CT and PET images, whereas for the survival network, the PET and CT image are fused by averaging.

For the localization model, the following random augmentations (with probability p) were applied the training samples: random intensity shifts in the range $(-0.5, 0.5)$ $(p = 0.5)$, random scale shifts with in the range $(-1, 1)$ for the PET and $(-0.25, 0.25)$ for the CT $(p = 0.5)$, and random Gaussian Noise with mean 0 and standard deviation 0.1 $(p = 0.1)$. The same augmentations reported in [10] were used to train the binary segmentation network. No augmentations were used for the survival prediction models.

3.3 Implementation Details

All models were run in a 32 GB NVidia Tesla V100, although they had different memory footprints and training times. The localization model was the lightest, whereas the binary segmentation model was the heaviest and with longer training time. All models were implemented using PyTorch [27], PyTorch Lighting [28] and PyTorch Geometric [29]. Table 3 summarizes the different hyperparameters and other training details for these three models.

Table 3. Implementation and training details of the different trained networks in this study.

	Detection	Segmentation	Survival (Deep Fusion)	Survival (Multi-patch)
Optimizer	Adam	SGD with momentum	Adam	Adam
Scheduler	Reduce LR on plateau	Poly LR	Multi step LR	Multi step LR
Initial learning rate	0.001	0.001	0.016	0.016
Loss function	Dice focal	Dice focal	MTLR	MTLR
Epoch	100	100	100	100
GPU RAM (GB)	4	26	10	10
Patch size	64^3	128^3	$80 \times 80 \times 50$	32^3
Training time (hours)	2	48	10	10
Validation metric	Average precision	Average precision	Concordance index	Concordance index

4 Results

Here we present our results, first for the segmentation task (Sect. 4.1), then for the survival prediction task (Sect. 4.2).

4.1 Segmentation Results

First, we assessed the binary segmentation and classification performance separately in the validation set. The binary segmentation network achieves a Dice score of 0.636, whereas for the tumour classification problem, the macro and micro F1 scores are 0.843 and 0.861 respectively.

Then, we obtained the semantic segmentation outputs from the binary segmentations and classifications results, and computed the aggregated Dice score. All the intersections are divided by all the unions in the considered data split independently for each class, and then the mean of the two is computed.

Table 4 reports this metric in the validation and test sets. Although the results suggest the proposed model is not a strong segmentor, it provides suitable input for the survival task, which benefits from the features extracted from the segmented tumours.

Table 4. Results of the proposed segmentation method in the validation and test sets.

Dataset split	GTVp Dice	GTVn Dice	Aggregated dice
Validation set (74 cases)	0.68514	0.62648	0.65581
Test set (359 cases)	0.59424	0.54988	0.57206

Finally, we qualitatively assessed the algorithm outputs by looking at the best, average and worst cases (Fig. 7). As a result of mistakes in the localization step, some of the predicted bounding box were slightly shifted from the actual RoI, which in turn resulted errors in the segmentation stage. For example, in *MDA-036* the predicted bounding box included a greater part of the brain, and the segmentation network detected a small region of brain as the tumour. Therefore, post-processing of the predicted bounding box could have improved the training stability and segmentation results.

4.2   Survival Prediction Results

Here we present our results for the survival prediction task. Table 5 reports the C-Index achieved by the proposed methods in the validation and test sets. For a fair comparison, we implemented and reported the results of *Deep Fusion*, the best performing neural network presented last year [25] for this task.

Table 5. Results of different methods in the validation set in terms of Concordance Index (C-Index). Best results in bold.

Model	Validation set (74 images)	Test set (339 images)
Weibull AFT	0.63014	**0.64086**
Deep Fusion [25]	0.60587	0.47923
Deep Fusion [25] + Weibull AFT	0.72194	0.64081
Multi-patch (ours)	**0.75000**	0.39679
Multi-patch + Weibull AFT (ours)	0.70536	0.64013

The proposed Multi-patch method performs best in the validation set, followed by Deep Fusion. Nevertheless, both deep learning methods generalize poorly to the test set, with our method obtaining the worst metric. To try to mitigate the overfitting, we ensembled the outputs of the deep learning and the Weibull AFT methods by simple averaging. However, even under this setting, the WeibulL AFT model alone was the best performing method in the test set. Little difference was observed between this model alone and ensembles of it and the deep learning algorithms.

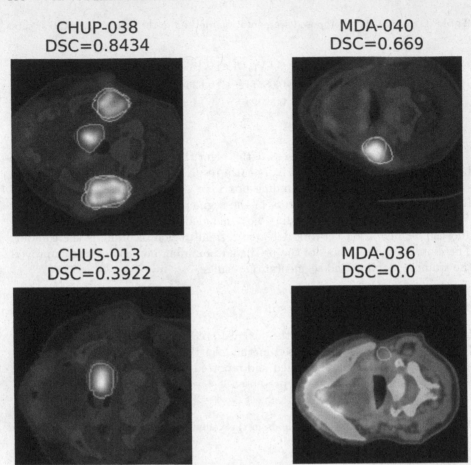

Fig. 7. Best, two average and worst segmentation outputs. The same preprocessing and averaging technique for the survival prediction is applied to the CT and PET images. Ground truth for GTVp and GTVn contours are in green and cyan, whereas predicted contours are in red and yellow, respectively. (Color figure online)

5   Discussion and Conclusion

We presented a framework for tumour segmentation and prognosis prediction in head & neck cancer patients, which may have a positive impact in patient management and personalized healthcare. Nevertheless, generalization still poses a challenge to the adoption of a solution based on neural networks. This has resulted in worse performance of the segmentation model, and even more so the survival model in the unseen cases of the test set.

The good generalization of Weibull AFT may due to the fewer parameters of this model compared with their deep learning counterparts, greatly reducing the possibility of overfitting. Some possible ways to mitigate this include reducing the neural network capacity (number of parameters), and to use *n-fold* cross validation and regularization during training. On the other hand, the superior performance of Weibull AFT with respect to Cox PH can be attributed to the acceleration/deceleration effect of the input features on the hazard probability, rather than their time independence, as assumed by the Cox PH model.

Finally, we have incorporated tumour-instance information in the prediction via processing tumour descriptors and tumour centred image patches with the improved graph attention networks [11]. Explanation algorithms like the approximated Shapley values [31] could be combined with the proposed method to increase the interpretability of predictions, a matter of crucial importance in clinical practice. Similar approaches have been used for gene expression data [32] and histopathology images [33]. We leave the application of these methods to head & neck cancer as future work.

References

1. Sung, H., et al.: Global cancer statistics 2020: GLOBOCAN estimates of incidence and mortality worldwide for 36 cancers in 185 countries. CA: A Cancer J. Clin. **71**(3), 209–249 (2021)

2. Johnson, D.E., Burtness, B., Leemans, C.R., et al.: Head and neck squamous cell carcinoma. Nat. Rev. Dis. Primers. **6**, 92 (2020)

3. Andrearczyk, V., et al.: Overview of the HECKTOR challenge at MICCAI 2020: automatic head and neck tumor segmentation in PET/CT. In: Andrearczyk, V., Oreiller, V., Depeursinge, A. (eds.) HECKTOR 2020. LNCS, vol. 12603, pp. 1–21. Springer, Cham (2021). https://doi.org/10.1007/978-3-030-67194-5_1

4. Oreiller, V., et al.: Head and neck tumor segmentation in PET/CT: the HECKTOR challenge. Med. Image Anal. **77**, 102336 (2022)

5. Andrearczyk, V., et al.: Overview of the HECKTOR challenge at MICCAI 2021: automatic head and neck tumor segmentation and outcome prediction in PET/CT images. In: Andrearczyk, V., Oreiller, V., Hatt, M., Depeursinge, A. (eds.) HECKTOR 2021. LNCS, vol. 13209, pp. 1–37. Springer, Cham (2021). https://doi.org/10.1007/978-3-030-98253-9_1

6. Andrearczyk, V., et al.: Overview of the HECKTOR challenge at MICCAI 2022: automatic head and neck tumor segmentation and outcome prediction in PET/CT. In: Andrearczyk, V., et al. (eds.) HECKTOR 2022. LNCS, vol. 13626, pp. 1–30. Springer, Cham (2023)

7. Ronneberger, O., Fischer, P., Brox, T.: U-Net: convolutional networks for biomedical image segmentation. In: Navab, N., Hornegger, J., Wells, W.M., Frangi, A.F. (eds.) MICCAI 2015. LNCS, vol. 9351, pp. 234–241. Springer, Cham (2015). https://doi.org/10.1007/978-3-319-24574-4_28

8. Isensee, F., et al.: nnU-Net: a self-configuring method for deep learning-based biomedical image segmentation. Nat. Methods **18**(2), 203–211 (2021)

9. Iantsen, A., Visvikis, D., Hatt, M.: Squeeze-and-excitation normalization for automated delineation of head and neck primary tumors in combined PET and CT images. In: Andrearczyk, V., Oreiller, V., Depeursinge, A. (eds.) HECKTOR 2020. LNCS, vol. 12603, pp. 37–43. Springer, Cham (2021). https://doi.org/10.1007/978-3-030-67194-5_4

10. Xie, J., Peng, Y.: The head and neck tumor segmentation based on 3D U-Net. In: Andrearczyk, V., Oreiller, V., Hatt, M., Depeursinge, A. (eds.) HECKTOR 2021. LNCS, vol. 13209, pp. 92–98. Springer, Cham (2022). https://doi.org/10.1007/978-3-030-98253-9_8

11. Brody, S., Alon, U., Yahav, E.: How attentive are graph attention networks? arXiv preprint arXiv:2105.14491 (2021)

12. Van der Walt, S., et al.: scikit-image: image processing in Python. PeerJ **2**(e453), e453 (2014)

13. Pedregosa, F., et al.: Scikit-learn: machine learning in Python. J. Mach. Learn. Res. **12**, 2825–2830 (2011)

14. Kalbfleisch, J.D., Prentice, R.L.: The Statistical Analysis of Failure Time Data. Wiley, New York (1980)

15. Cox, D.R.: Regression models and life-tables. J. Roy. Stat. Soc.: Ser. B (Methodol.) **34**(2), 187–202 (1972). https://doi.org/10.1111/j.2517-6161.1972.tb00899.x

16. Yu, C.-N., et al.: Learning patient-specific cancer survival distributions as a sequence of dependent regressors. In: Advances in Neural Information Processing Systems, vol. 24 (2011)

17. Jin, P.: Using survival prediction techniques to learn consumer-specific reservation price distributions. University of Alberta, Edmonton, AB (2015)

18. Fotso, S., et al.: Deep neural networks for survival analysis based on a multi-task framework. arXiv:1801.05512

19. Ioffe, S., Szegedy, C.: Batch normalization: accelerating deep network training by reducing internal covariate shift. In: International Conference on Machine Learning, pp. 448–456. PMLR, June 2015

20. Milletari, F., et al.: V-Net: fully convolutional neural networks for volumetric medical image segmentation. In: 2016 Fourth International Conference on 3D Vision (3DV). IEEE (2016)

21. Ulyanov, D., Vedaldi, A., Lempitsky, V.: Instance normalization: the missing ingredient for fast stylization. arXiv preprint arXiv:1607.08022 (2016)

22. Lin, T.-Y., et al.: Focal loss for dense object detection. In: Proceedings of the IEEE ICCV (2017)

23. Sudre, C.H., Li, W., Vercauteren, T., Ourselin, S., Jorge Cardoso, M.: Generalised dice overlap as a deep learning loss function for highly unbalanced segmentations. In: Cardoso, M.J., et al. (eds.) DLMIA/ML-CDS -2017. LNCS, vol. 10553, pp. 240–248. Springer, Cham (2017). https://doi.org/10.1007/978-3-319-67558-9_28

24. Davidson-Pilon, C., et al.: Lifelines: survival analysis in Python. J. Open Source Softw. **4**(40), 1317 (2019). https://doi.org/10.21105/joss.01317

25. Saeed, N., Al Majzoub, R., Sobirov, I., Yaqub, M.: An ensemble approach for patient prognosis of head and neck tumor using multimodal data. In: Andrearczyk, V., Oreiller, V., Hatt, M., Depeursinge, A. (eds.) HECKTOR 2021. LNCS, vol. 13209, pp. 278–286. Springer, Cham (2022). https://doi.org/10.1007/978-3-030-98253-9_26

26. Akiba, T., Sano, S., Yanase, T., Ohta, T., Koyama, M.: Optuna: a next-generation hyperparameter optimization framework. CoRR abs/1907.10902 (2019)

27. Paszke, A., et al.: PyTorch: an imperative style, high-performance deep learning library. In: Advances in Neural Information Processing Systems, vol. 32 (2019)

28. Falcon, W., et al.: PyTorch lightning (2019). https://doi.org/10.5281/zenodo.3828935

29. Fey, M., et al.: Fast graph representation learning with PyTorch geometric. arXiv preprint arXiv:1903.02428 (2019)

30. Fatima, S.S., et al.: A linear approximation method for the Shapley value. Artif. Intell. **172**(14), 1673–1699 (2008)
31. Ancona, M., et al.: Explaining deep neural networks with a polynomial time algorithm for shapley value approximation. In: ICML, pp. 272–281. PMLR, May 2019
32. Hayakawa, J., et al.: Pathway importance by graph convolutional network and Shapley additive explanations in gene expression phenotype of diffuse large B-cell lymphoma. PLoS ONE **17**(6), e0269570 (2022)
33. Bhattacharjee, S., et al.: An explainable computer vision in histopathology: techniques for interpreting black box model. In: International Conference on Artificial Intelligence in Information and Communication (ICAIIC). IEEE (2022)

Combining nnUNet and AutoML for Automatic Head and Neck Tumor Segmentation and Recurrence-Free Survival Prediction in PET/CT Images

Qing Lyu[✉][iD]

Faculty of Art and Science, University of Toronto, Toronto, Canada
amy.lyu@mail.utoronto.ca

Abstract. Head and neck (H&N) tumor segmentation from FDG-PET/CT images has a significant impact on radiotherapy diagnosis. However, manual delineation of primary tumor and lymph nodes is a time-consuming and labor-intensive process. Also, patients that underwent radiotherapy have a high risk of regional recurrence. In this work, we used 3D nnU-Net with DiceTopK loss function to achieve automatic segmentation for head and neck primary tumor and lymph nodes. With the average ensemble of five cross-validation models, our approach got an aggregated Dice Similarity Coefficient of 0.70427 on the test set. Furthermore, we extracted radiomics features from PET/CT images combined with the clinical data to predict patients' Recurrence-Free Survival (RFS) using a weighted ensemble predictor by AutoGluon. This led to a Concordance index of 0.63896 on the test set. The code is publicly available at https://github.com/amylyu1123/HECKTOR-2022.

Keywords: Head and neck cancer · 3D U-Net · DiceTopK loss · AutoGluon · Recurrence-free survival

1   Introduction

One of the most common malignant cancers is head and neck (H&N) cancer [8]. Radiation therapy, also called radiotherapy, is one of the primary curative treatments for head and neck cancer [1]. The fluorodeoxyglucose PET (FDG-PET) and CT scans of the head and neck primary tumor (GTVp) have a significant impact on the radiotherapy diagnosis and planning [7,14]. Simultaneously, it is crucial to detect the lymph nodes (GTVn) that are close to the primary tumor for head and neck cancers. Otherwise, inappropriate treatment will be carried out for the patient, which will lead to the metastasis of cancer or even death [13]. However, accurate detection and the manual delineation of both the primary tumor and lymph nodes of head and neck cancer is time-consuming and labor-intensive process. An automatic process that provides the tumor and lymph nodes segmentation from FDG-PET/CT with high accuracy and efficiency is

V. Andrearczyk et al. (Eds.): HECKTOR 2022, LNCS 13626, pp. 192–201, 2023.
https://doi.org/10.1007/978-3-031-27420-6_19

vital. In addition, the high percentage of loco-regional recurrence of head and neck cancers after radiotherapy has become a problem [1]. Some of the clinical information and tumor conditions of the patient are prognostic factors that determine Recurrence-Free Survival (RFS) for head and neck cancers [2]. The process of choosing these prognostic factors that are related to RFS is time-consuming. This leads to the importance of constructing an automatic model to predict and analyze the patient's RFS.

The field of deep learning, especially convolutional neural networks (CNN), is commonly used for automatic image segmentation in medical image analysis [16]. Using automated machine learning that combines multiple models through ensembling can be an effective approach to achieve Recurrence-Free Survival prediction with high accuracy.

In this paper, we tried experimenting with two tasks proposed by the HECK-TOR 2022 challenge, the HEad and neCK TumOR segmentation and outcome prediction in PET/CT images [4,12]. Task 1 is to develop automatic segmentations of Head and Neck (H&N) primary tumor (GTVp) and lymph nodes (GTVn) based on 3D bi-modal FDG-PET/CT images. We used the state-of-art method, 3D nnU-Net [9], to achieve the automatic supervised segmentation by extending it with DiceTopK as its loss function. Task 2 is to predict patients' Recurrence-Free Survival (RFS) using the PET/CT images and clinical data. We tried AutoGluon [6], particularly AutoGluon-Tabular. It is an AutoML framework which solves the regression problem by ensembling various models, including state-of-art machine learning models and deep learning models, and stacking them in multiple layers. The weighted ensemble model that has a stack level of 2 attains the best predictive performance in order to predict the RFS.

2 Method

2.1 Preprocessing

For Task 1, at the beginning of the preprocessing, all images were cropped by only including the non-zero regions. As the data was anisotropic with a target spacing of $4 \times 3.27 \times 4\,\mathrm{mm}$, the PET/CT images were resampled with the nearest neighbor interpolation along the out-of-plane axis and the third-order spline interpolation along the in-plane axis. The annotations were converted into one-hot at first. Then, they were resampled with linear interpolation along the in-plane axis. All decimal numbers were converted into integers by using a threshold of 0.5. The nearest neighbor interpolation was used along the out-of-plane axis of the annotations. Apart from that, the intensity of PET image was normalized using the Z-score, which was subtracting the mean and dividing it by the standard deviation. For CT images, they were first normalized by clipping to the 0.5 and 99.5 percentiles of the foreground voxels. Then, they were normalized by the global foreground mean and standard deviation.

For Task 2, at first, we merged the primary tumor and lymph nodes together using the same label. Then, the radiomics features were extracted from both PET and CT images with the mask using the Pyradiomics package [15]. We used the default setting when set up the parameters and instantiated the extractor. The extracted radiomic features[1] included these classes: first order statistics, shape-based (3D), shape-based (2D), Gray Level Co-occurrence Matrix (GLCM), Gray Level Run Length Matrix (GLRLM), Gray Level Size Zone Matrix (GLSZM), Neighbouring Gray Tone Difference Matrix and Gray Level Dependence Matrix (GLDM). For shape-based features that have repeated values extracted from the PET and CT image, only one of them was kept. This led to a total number of 207 features. A logarithmic operation was applied to the features that had large values. For the clinical information, the gender was converted to a one-hot format. The features such as weight, tobacco, alcohol, performance status, HPV status and surgery were dropped as many of the observations had missing values. Furthermore, the feature of center ID was dropped as this may not be correlated with the outcome. Generally in practice, our model should be able to predict using different datasets which may not contain the center ID. Keeping it in the model would lead to bias, so we did not include it in the model. The clinical features we kept were age, gender, chemotherapy. Finally, the training set that was provided was split into two new subsets, where 80% of them formed the new training set and the rest 20% were the validation set.

2.2 Proposed Method

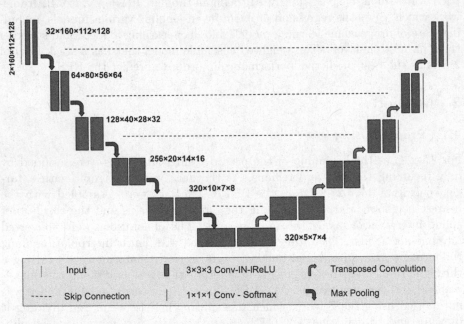

Fig. 1. 3D nnU-Net architecture.

For Task 1, inspired by the top approaches from the previous HECKTOR 2021 challenge [3], we also used the nnU-Net [9] as the main network, which auto-

---

[1] https://pyradiomics.readthedocs.io/en/latest/features.html.

matically configures itself and requires no expert knowledge for the automatic segmentation of the primary tumor and lymph nodes.

Specifically, the 3D U-Net on the full image resolution, shown in Fig. 1, was the network we used. According to the winner's approach from previous HECK-TOR 2021 challenge, we concatenated CT and PET images into a two-channel tensor as input [3]. The input size was $2 \times 160 \times 112 \times 128$. For each resolution state of the encoder and decoder, there were two computational blocks. It included a convolution, instance normalization and leaky ReLU (conv - instance norm - leaky ReLU). For downsampling, strided convolutions were applied. Transposed convolutions were carried out for upsampling. The number of feature map at the initial point was 32. It doubled after one downsampling and the maximum is 320. The stride of the max pooling operation was (2, 2, 2), except for the last one, which was (2, 1, 2).

The default loss function used by nnU-Net is the compound loss of Dice and cross-entropy loss. In order to improve the performance of nnU-Net on image segmentation, we extended it by using a new loss function: the combination of Dice and TopK loss.

The TopK loss function is defined by

$$L_{TopK} = -\frac{1}{N} \sum_{c=1}^{C} \sum_{i \in K} g_i^c logs_i^c$$

where g_i and s_i denote the ground truth and segmentation of voxel i respectively, C denotes the number of classes, and K denotes the set containing the worst $k\%$ voxels. TopK loss is similar to the cross-entropy loss, except for a percentage k would be set, which allows networks to pay attention to the hard samples [11]. We used $k = 10$ for network training, which means that only the top 10% worst pixels would be considered when computing the loss function because the TopK-10% loss has the best performance on challenging segmentation tasks [11]. Since the compound loss functions have been shown as the most robust loss in 3D image segmentation, then the summation of Dice and TopK loss was used as the loss function for network training. In addition, five-fold cross-validation was performed on the training set. In the end, the average ensemble of the five models was used to conduct the inference on the test set.

For Task 2, we used the AutoGluon [6] on Tabular Data, which can automatically apply ensembling from machine learning and deep learning models to achieve high accuracy and predictive performance. Although AutoGluon is designed for supervised learning tasks, classification and regression, predicting Recurrence-Free Survival can be considered as a regression task.

Figure 2 demonstrates the ensemble method with the stacking strategy used in AutoGluon. Unlike common ensemble methods, AutoGluon's ensemble method consists of stacking with multiple layers [6]. Level 1 used multiple base models, depicted by model 1 to model N, to make predictions using input data. Their predictions were concatenated with the input data as the input for the second level. Their predictions were weighted as the final output. This architecture was similar to the neural network model which consists of multiple layers. Thus, this kind of multi-layer stack ensembling utilized the features optimally.

Instead of using AutoGluon's default setting in the fit function, we chose to change some configurations for model training. The mean absolute percentage error was used as the evaluation metric on the validation set. The preset was set to best quality to attain models with high predictive accuracy. The auto stack was set to True, which allowed AutoGluon to automatically choose good values for ensemble with bagging or stacking. Models were trained on the new training set split from 80% of the original training set.

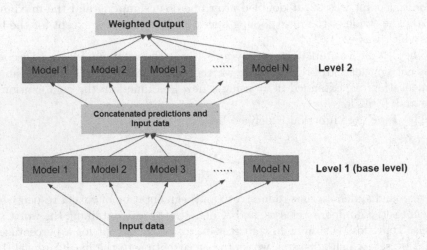

Fig. 2. The multi-layer (2 stacking layers) stack ensemble method for AutoGluon.

2.3   Post-processing

For Task 1, the segmentation of GTVp and GTVn were resampled into the same dimensions and spacings of the original CT images using linear interpolation. For Task 2, all predictions of the Recurrence-Free Survival in days were multiplied by −1 to have predicted risk scores.

3   Experiments

3.1   Dataset and Evaluation Measures

The dataset used in this paper comes from the HECKTOR 2022 challenge, which is the Third Edition of the Head and Neck Tumor Segmentation and outcome prediction in PET/CT images [4,12].

For Task 1, the training set includes 524 patients with oropharyngeal head and neck cancer, coming from 7 different clinical centers. For each patient, the

FDG-PET and low-dose non-contrast-enhanced CT images of the head and neck region are included. The patient's primary Gross Tumor Volumes (GTVp) and nodal Gross Tumor Volumes (GTVn) are labelled in one NIfTI image as well. The test set has 359 cases, including the PET/CT image data and the annotation of the primary tumor (GTVp) and lymph nodes (GTVn).

For Task 2, the number of patients in the training set and test set is 489 and 339, respectively. Their PET/CT images come from Task 1, excluding the individuals with incomplete responses. Meanwhile, the clinical information, survival events and Recurrence-Free Survival are given. Recurrence-Free Survival represents the length of time (number of days) between the end of the patient's radiotherapy and the events or last follow-up. This includes events such as death, local, regional and distant mets. During this period, patients all live without cancer.

For Task 1, the evaluation metric used in this experiment is the aggregated Dice Similarity Coefficient (aggregated DSC), which is adapted from the Aggregated Jaccard Index [10]. Aggregated DSC is denoted by the equation below. The ranking will be based on the average of the aggregated DSC of GTVp and GTVn as both of them are equally important.

$$DSC_{agg} = \frac{2\sum_i^N \sum_k g_{i,k} s_{i,k}}{\sum_i^N \sum_k (g_{i,k} + s_{i,k})}$$

where N denotes the size of the test set, $g_{i,k}$ and $s_{i,k}$ denote the ground truth and segmentation on image i and voxel k respectively.

For Task 2, the Concordance index (C-index) will be used to evaluate the model's ability by individual's risk score on the test data.

3.2   Implementation Details

Environment Settings. Table 1 demonstrates our development environments and requirements for Task 1.

Table 1. Development environments and requirements for Task 1.

Operating system	Linux - CentOS 7
RAM	60 G
CPU	Intel(R) Xeon(R) CPU E5-2683 v4 @ 2.10 GHz
GPU (number and type)	V100-SXM2 32 G
CUDA version	11.6
Programming language	Python 3.8
Deep learning framework	Pytorch (Torch 1.10, torchvision 0.11.1)

Table 2. Training protocols for Task 1.

Batch size	2
Patch size	$160 \times 112 \times 128$
Total epochs	1000
Optimizer	SGD with nesterov momentum ($\mu = 0.99$)
Initial learning rate (lr)	0.01
Training time	6 days

Training Protocols. Table 2 shows the training protocols used in our network for Task 1.

4   Results and Discussion

4.1   Quantitative Results on Validation Set

For Task 1, the five-fold cross-validation was conducted according to our proposed method. Table 3 shows the aggregated DSC for the primary tumor (GTVp) and lymph nodes (GTVn) for each fold respectively on the training set. The last column is the average aggregated DSC between GTVp and GTVn. Fold 1 attained the highest aggregated DSC for both GTVp and GTVn. The aggregated DSC of GTVp and GTVn for all other folds were between 0.60 and 0.68.

Table 3. Aggregated DSC of GTVp and GTVn using five-fold cross-validation on the training set.

Fold number	GTVp DSC_{agg}	GTVn DSC_{agg}	Average DSC_{agg}
0	0.6639	0.6731	0.6685
1	0.7039	0.7087	0.7063
2	0.6797	0.6666	0.6732
3	0.6394	0.6089	0.6241
4	0.6518	0.6373	0.6446
Average	0.6677	0.6589	0.6633

For Task 2, 12 models were used to fit the training set by AutoGluon. Details can be found in Appendix. The best model was the weighted ensemble with a stack level of 2 (WeightedEnsemble_L2) [5]. We got the highest value of the C-index by using the parameter configuration in the proposed method with our best model, which was 0.65453 on the validation set stated in the preprocessing.

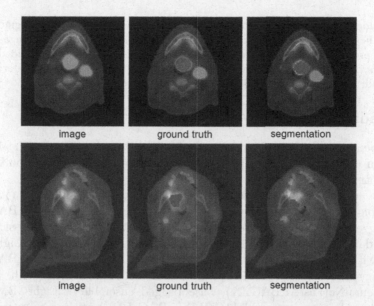

Fig. 3. For each row (from left to right): the PET (use hot color map) and CT image displayed as one layer; PET/CT with the ground truth of GTVp (red) and GTVn (green); PET/CT with the segmentation of GTVp and GTVn. First row is a good example of segmentation. Second row is a bad example. (Color figure online)

4.2 Qualitative Results on Validation Set for Task 1

Figure 3 shows a good and bad example of segmentation from the training set. The good example attained aggregated DSC of 0.8916 and 0.8010 for GTVp and GTVn respectively. The bad example attained aggregated DSC of 0.3224 and 0.5006 for GTVp and GTVn respectively.

5 Conclusion

For Task 1, by predicting the segmentation on the test set using the ensemble of the five models from the cross-validation, we got an average aggregated DSC of 0.7042 on the test set. For Task 2, we used our predicted segmentation as the label to extract radiomics features. By predicting RFS using the best configuration of the function parameters, we got a C-index of 0.63896 on the test set.

In conclusion, the 3D nnU-Net can produce automatic segmentation of the head and neck tumor and lymph nodes with high efficiency. In future work, we can extend the nnU-Net in other ways in order to achieve better segmentation results. In addition, AutoML is a good and efficient way to fit models chosen from multiple machine learning models based on different datasets. In future work, we can try some hyperparameter tuning on specific models to improve their performance.

Acknowledgements. The author of this paper declares that the segmentation method they implemented for participation in the HECKTOR 2022 challenge has not used any pre-trained models nor additional datasets other than those provided by the organizers. The proposed solution is fully automatic without any manual intervention. The author would like to thank Jun Ma for useful discussions.

Appendix

This appendix lists 12 fitted models used for Task 2 by AutoGluon. The detailed definition can be found at https://auto.gluon.ai/stable/api/autogluon.tabular. models.html.

They are the weighted ensemble models with a stack level of 2 (WeightedEnsemble_L2), base bagged ensemble of XGBoost models (XGBoost_BAG_L1), base bagged ensemble of PyTorch neural network models for regression with tabular data (NeuralNetTorch_BAG_L1), base bagged ensemble of LightGBM models with specific hyperparameter configuration (LightGBMLarge_BAG_L1), base bagged ensemble of fastai v1 neural network models with tabular data (NeuralNetFastAI_BAG_L1), base bagged ensemble of LightGBM models that enable extra trees (LightGBMXT_BAG_L1), base bagged ensemble of LightGBM models (LightGBM_BAG_L1), base bagged ensemble of CatBoost models (CatBoost_BAG_L1), base bagged ensemble of random forest models (RandomForestMSE_BAG_L1), base bagged ensemble of Extra Trees models (ExtraTreesMSE_BAG_L1), base bagged ensemble of KNearestNeighbors models with uniform weights (KNeighborsUnif_BAG_L1) and base bagged ensemble of KNearestNeighbors models with weights set to distance (KNeighborsDist_BAG_L1).

References

1. Alterio, D., Marvaso, G., Ferrari, A., Volpe, S., Orecchia, R., Jereczek-Fossa, B.A.: Modern radiotherapy for head and neck cancer. Semin. Oncol. **46**(3), 233–245 (2019)
2. Andersson, A.P., Gottlieb, J., Drzewiecki, K.T., Hou-Jensen, K., Sondergaard, K.: Skin melanoma of the head and neck: prognostic factors and recurrence-free survival in 512 patients. Cancer **69**(5), 1153–1156 (2010)
3. Andrearczyk, V., et al.: Overview of the HECKTOR challenge at MICCAI 2021: automatic head and neck tumor segmentation and outcome prediction in PET/CT images. In: Andrearczyk, V., Oreiller, V., Hatt, M., Depeursinge, A. (eds.) Head and Neck Tumor Segmentation and Outcome Prediction, HECKTOR 2021. Lecture Notes in Computer Science, vol. 13209, pp. 1–37. Springer, Cham (2022). https://doi.org/10.1007/978-3-030-98253-9_1
4. Andrearczyk, V., et al.: Overview of the HECKTOR challenge at MICCAI 2022: automatic head and neck tumor segmentation and outcome prediction in PET/CT. In: Head and Neck Tumor Segmentation and Outcome Prediction (2023)
5. Caruana, R., Niculescu-Mizil, A., Crew, G., Ksikes, A.: Ensemble selection from libraries of models. In: Proceedings of the Twenty-First International Conference on Machine Learning, ICML 2004, p. 18. Association for Computing Machinery, New York, NY, USA (2004)

6. Erickson, N., et al.: Autogluon-Tabular: Robust and Accurate AutoML for Structured Data. arXiv preprint arXiv:2003.06505 (2020)
7. Escott, E.J.: Role of positron emission tomography/computed tomography (PET/CT) in head and neck cancer. Radiol. Clin. North Am. **51**(5), 881–893 (2013)
8. Farris, C., Petitte, D.M.: Head, neck, and oral cancer update. Home Healthc. Nurse **31**(6), 322–328 (2013)
9. Isensee, F., Jaeger, P.F., Kohl, S.A.A., et al.: nnU-Net: a self-configuring method for deep learning-based biomedical image segmentation. Nat. Methods **18**, 203–211 (2021)
10. Kumar, N., Verma, R., Sharma, S., Bhargava, S., Vahadane, A., Sethi, A.: A dataset and a technique for generalized nuclear segmentation for computational pathology. IEEE Trans. Med. Imaging **36**(7), 1550–1560 (2017)
11. Ma, J., et al.: Loss odyssey in medical image segmentation. Med. Image Anal. **71**, 102035 (2021)
12. Oreiller, V., et al.: Head and neck tumor segmentation in PET/CT: The HECK-TOR challenge. Med. Image Anal. **77**, 102336 (2022)
13. Roepman, P., Wessels, L., Kettelarij, N., et al.: An expression profile for diagnosis of lymph node metastases from primary head and neck squamous cell carcinomas. Nat. Genet. **37**, 182–186 (2005)
14. Send, T., Kreppel, B., Gaertner, F.C., et al.: PET-CT bei Karzinomen im KopfHalsBereich. HNO **65**, 504–513 (2017)
15. van Griethuysen, J.J., et al.: Computational radiomics system to decode the radiographic phenotype. Cancer Res. **77**(21), e104–e107 (2017)
16. Wang, X., Li, B.b.: Deep Learning in head and neck tumor multiomics diagnosis and analysis: review of the literature. Front. Genet. **12**, 624820 (2021)

Head and Neck Cancer Localization with Retina Unet for Automated Segmentation and Time-To-Event Prognosis from PET/CT Images

Yiling Wang[1,2], Elia Lombardo[1], Lili Huang[1], Claus Belka[1,3], Marco Riboldi[4], Christopher Kurz[1], and Guillaume Landry[1(✉)]

[1] Department of Radiation Oncology, University Hospital, LMU Munich, Munich, Germany
guillaume.landry@med.uni-muenchen.de
[2] Department of Radiation Oncology, Radiation Oncology Key Laboratory of Sichuan Province, Sichuan Clinical Research Center for Cancer, Sichuan Cancer Hospital & Institute, Sichuan Cancer Center, Affiliated Cancer Hospital of University of Electronic Science and Technology of China, Chengdu, China
[3] German Cancer Consortium (DKTK), Partner Site Munich, Munich, Germany
[4] Department of Medical Physics, Ludwig-Maximilians-Universität München, Garching, Germany

Abstract. Auto-segmentation of the primary tumor (GTVp) and the associated lymph nodes (GTVn) of head and neck cancer (HNC) is deemed beneficial for precise radiotherapy. However, previous studies were mostly focusing on the auto segmentation of GTVp. It remains difficult to additionally segment GTVn automatically. Moreover, current methods also face challenges in handling whole-body scans due to memory limitations. In this study, the Retina Unet has been utilized for the first time to localize the HNC from whole-body positron emission tomography/computed tomography (PET/CT) scans. Cropped (based on the predicted tumor center) PET/CTs were then input to a multi-label Unet to produce the auto segmentation of GTVp and GTVn. Several time-to-event analysis models, including a segmentation-free model, have also been explored for relapse free survival (RFS) prognosis. The proposed models achieved encouraging cross validation (testing) Dice coefficient scores for GTVp/GTVn segmentation of 0.66 (0.70), indicating a promising first step towards fully automated HNC localization and segmentation from whole-body PET/CT scans. Furthermore, the segmentation-free PET-only RFS prognosis model produced the best average cross-validation (testing) Harrell's Concordance Index of 0.70 (0.635), verifying our previous observation that GTV segmentation might be less relevant for PET-based prognosis.

Keywords: Head and neck cancer · PET/CT · Tumor localization · Auto-segmentation · Outcome prediction

---

C. Kurz and G. Landry—Equally contributing authors.

V. Andrearczyk et al. (Eds.): HECKTOR 2022, LNCS 13626, pp. 202–211, 2023.
https://doi.org/10.1007/978-3-031-27420-6_20

1 Introduction

Head and neck cancer (HNC), which is the sixth most frequently occurring cancer worldwide [1], is conventionally treated with radiotherapy, chemotherapy, surgery, or combined modalities [2]. However, due to the development of distant metastasis or second primary cancers, the long-term survival of HNC can be as low as 50%, while up to 40% of the patients may suffer a loco-regional failure in the first two years after treatment [3, 4]. Therefore, accurate identification of the gross tumor volume (GTV) and classification of high-risk patients before treatment remain crucial challenges in HNC treatment.

Recently, several radiomics studies, especially those based on [18F] fluorodeoxyglucose (FDG) positron emission tomography (PET), have shown promising capability for HNC prognosis in a non-invasive manner [5, 6]. Compared to anatomical images, such as computed tomography (CT) and magnetic resonance imaging (MRI), PET may detect hypoxia levels [7] and reflect the physiological changes related to tumor cellular metabolism, thus serving as a more relevant image biomarker for prognosis. Besides, the recent development of deep learning methods has enabled auto-segmentation of primary tumor (GTVp), a competitive alternative to avoid the time-consuming and error-prone manual delineation, which has been validated by the previous HECKTOR challenges in 2020 and 2021 [8, 9, 15]. However, the previous deep learning studies seldom dealt with the associated lymph nodes (GTVn) in HNC, which also carry nontrivial information on the expected outcome via the presence of metastatic lesions. Additionally, although it is quite common to have whole-body PET/CT scans in clinics, previous studies are mostly based on a manually selected bounding boxes, while auto-localization of HNC GTVs from entire images remains challenging.

To face the new HECKTOR challenge in 2022, we used Retina Unet, a deep learning network for tumor localization [10], to find HNC in whole-body PET/CT scans, and later utilized a multi-label Unet to segment both GTVp and GTVn. Moreover, to further explore the predictive potential of PET/CT, we extended our recent segmentation-free HNC time-to-event analysis model [11] to the prognosis of recurrence-free survival (RFS). The goal of this study was thus to explore the feasibility of whole-body PET/CT localization and segmentation for both GTVp and GTVn and to implement time-to-event RFS prognosis in HNC patients.

2 Methods

2.1 Dataset

As provided by HECKTOR 2022, nine different cohorts with a total of 524 histologically proven oropharyngeal HNC patients were used for training the segmentation task, while 489 cases were used for the RFS prognosis task. All patients underwent radiotherapy and/or chemotherapy treatment, had FDG-PET as well as low-dose non-contrast-enhanced CT images, and had GTVp and GTVn contoured and attributed to labels 1 and 2 as segmentation masks, respectively. Besides, the PET intensities had already been converted to standardized uptake values and the CT images had been aligned with the corresponding PET. It was noted that several patients had whole-body PET/CT scans.

For the prognosis task, the clinical information of each patient was provided. The survival times and the times between the end of radiotherapy and an event or last follow-up were also provided in days. Since only those patients who had complete responses were selected for the prognosis task, a smaller number of patients was provided compared with the segmentation task.

For independent testing, another set of 359 patients was provided by the HECKTOR organizers for the segmentation task, including 339 also for the prognosis task.

2.2 Tumor Localization and Image Preprocessing

We adopted the Retina Unet [10] to localize the tumor center from whole-body PET/CTs. The input PET/CTs and their segmentation masks were firstly cropped (from the axial image center) to a size of 512×512 mm in the axial plane with 2 mm pixel grid spacing and were resampled to 3 mm grid spacing along the superior-inferior direction. The resampling for PET and CT was implemented with linear interpolation, whereas the resampling for segmentation masks relied on nearest-neighbour interpolation throughout this study. For the retina Unet, GTVp and GTVn were combined as a single label to find the tumor (i.e., the combination of GTVp and GTVn) center. To train the network, a multi-task loss function was applied

$$L_{retina\_unet} = L_c + L_b + L_s \tag{1}$$

where L_c was the class loss defined as Eq. (5) in [16], L_b was the bounding box loss defined as Eq. (3) in [17], and L_s was the segmentation loss defined as a combination of soft dice loss L_d and the pixel-wise cross-entropy loss in Eq. (1) of [10]. L_d can stabilize the class-imbalance problem in segmentation tasks and was formulated as

$$L_d = -\frac{2}{|K|} \sum_{k \in K} \frac{\sum_{i \in I} u_{i,k} v_{i,k}}{\sum_{i \in I} u_{i,k} + \sum_{i \in I} v_{i,k}} \tag{2}$$

with u the softmax output of the network and v the one hot encoding of the ground truth segmentation map. Both u and v had shape $I \times K$ with $i \in I$ being the number of pixels in the training batch and $k \in K$ being the classes.

A dual channel (with PET and CT as inputs) six-layer Retina Unet [10] with ResNet50 [12] as the backbone was trained in a slice-based (2D) manner to determine the center of the tumor in each slice. These slice centers were later ensembled to localize the center of the tumor in 3D. In this study, the Retina Unet was trained with 80% of the provided training dataset (419 patients) and validated on the remaining 20% (105 patients). It was trained for 100 epochs using the Adam optimizer (learning rate 1e-3) on an NVIDIA Quadro RTX 8000 (48 GB) GPU with a batch size of 50. To avoid overfitting, data augmentation was applied on the fly during training, including rotation, elastic deformation, and random scaling. More details of the network architecture can be found in [10].

In our approach, the Retina Unet outputs both the coordinates of a tumor bounding box and the corresponding confidence score (ranging from 0 to 1). Since the GTVp and GTVn can be distinct volumes, there could be several predicted boxes for each slice.

To determine the center of the combined GTVp and GTVn, we computed the true and predicted box center differences in terms of confidence score thresholds ranging from 0.4 to 0.9, effectively treating the threshold as a hyper-parameter optimized based on the validation set. If a box had a confidence score larger than the threshold times the maximum confidence score of that patient, then it would be collected to compute the tumor volume center. Finally, we took the median center coordinates in 3D of those collected boxes as the center for the tumor volume.

Based on the tumor center coordinates determined from the Retina Unet, we later cropped and resampled the PET/CTs and their segmentation masks to a volume size of $256 \times 256 \times 256$ mm on a 1 mm isotropic grid. For the segmentation task, due to a limitation of memory, the input images (PET/CT and segmentation mask) were resampled to $144 \times 144 \times 144$ voxels with 1.78 mm grid spacing. For the prognosis task, the input images were kept as $256 \times 256 \times 256$ voxels with 1.0 mm grid spacing. In addition, the z-score normalization was implemented for PET/CT images in both models.

2.3 Auto-segmentation

We used a dual channel 4-level Unet [13] to segment GTVp (as label 1) and GTVn (as label 2) automatically from the cropped PET/CT based on the tumor center determined by Retina Unet. The input images (PET and CT) were both $144 \times 144 \times 144$ in pixel size. To address class imbalance, the default soft dice loss function L_d defined in Eq. (2) was applied. Besides, we also combined multi-label cross-entropy loss with the soft dice loss function with equal weight to compare the segmentation performance.

All 524 patients were used for training. To obtain the optimal hyper-parameters, we applied 3-fold and 5-fold cross-validation (CV), resulting in three or five trained models with the highest average dice score for GTVp and GTVn. To get the segmentation masks for the testing dataset, the mean value of the predicted probability maps (GTVp and GTVn separately) from all the CV folds was computed and thresholded at 0.5 to get the respective label masks.

The network was trained for 300 epochs using the Adam optimizer on an NVIDIA Quadro RTX 8000 (48 GB) GPU with a batch size of 4. The initial learning rate was 5e-4 and decayed at a rate of 2 if the validation loss was not improved after every 20 epochs. To avoid overfitting, data augmentation was applied including random flipping, random shifting, and mirroring.

2.4 RFS Prognosis

We used the 3D Resnet from our recent study [11] to perform RFS time-to-event prognosis and investigated performance changes with and without segmentation masks. The corresponding inputs of the network were PET-only (PET without segmentation mask), PET/Seg (PET and segmentation mask as separate input channels), CT/Seg (CT and segmentation mask as separate input channels), and PET/CT/Seg (PET, CT and segmentation mask as separate input channels), respectively. The segmentation mask was predicted from the multi-label Unet with label 1 for GTVp and label 2 for GTVn. Our time-to-event model predicts RFS probabilities at up to 5 years with a spacing of half a year and the predictive performance was evaluated by Harrell's concordance index (HCI) after 3 years. Accordingly, we used the custom survival likelihood cost function defined in Eq. (2) by Gensheimer et al. [11].

To obtain the optimal hyper-parameters, we applied 3-fold CV, resulting in three trained models which achieved the best average CV performance. The training dataset was split into stratified folds with StratifiedKFold (sklearn package) during the CV process to preserve the class ratio of RFS. To get the final result for the testing cohort, we averaged the predictions of the three models to obtain a single model-averaged HCI.

The network was trained for 500 epochs using the Adam optimizer on an NVIDIA RTX A6000 (48 GB) and the batch size was 8 for single channel input and 4 for dual-channel input. To improve the training efficiency, an early stopping with a patience of 200 epochs was also implemented. The data augmentation and learning rate schemes were kept the same as recommended in Table 2 of reference [11].

3　Results

3.1　Tumor Localization with Retina Unet

The obtained maximum 3D tumor center differences with confidence score thresholds ranging from 0.4 to 0.9 are summarized in Table 1. For the validation cohort, the difference was less than 40 mm in each direction with the optimal threshold of 0.6. Figure 1 shows the histogram of tumor center differences for all patients (training and validation). It shows that a higher confidence score threshold, while leading to precise localization for most patients, may lead to outliers. Therefore, to avoid outliers, we chose the confidence score threshold as 0.6 in this study, accepting less precise but more robust localization.

Table 1. Maximum tumor center differences in superior-inferior, lateral and anterior-posterior directions

Threshold	Training dataset			Validation dataset		
	Superior-inferior	Lateral	Anterior-posterior	Superior-inferior	Lateral	Anterior-posterior
	(mm)					
0.4	65	46	32	33	54	29
0.5	60	46	32	32	57	29
0.6	60	47	31	30	37	29
0.7	56	47	32	180	40	29
0.8	55	49	41	180	58	29
0.9	54	49	33	180	58	55

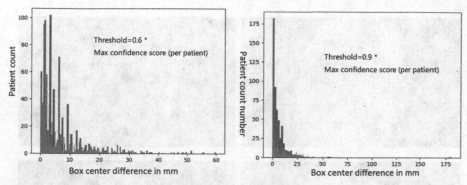

Fig. 1. Histogram of tumor center differences (in mm) with thresholds 0.6 and 0.9. The values in superior-inferior, lateral and anterior-posterior directions are shown in blue, orange and green, respectively. The x-axis denotes the box center difference in mm, and the y-axis denotes the patients count for both training and validation cohorts (in a total of 524 patients) (Color figure online)

3.2 Auto-segmentation Task

The dice coefficient (DSC) was computed to evaluate the segmentation performance. For each patient, it was calculated as

$$DSC = \frac{2\sum_{i \in I} t_i p_i}{\sum_{i \in I} t_i + \sum_{i \in I} p_i} \tag{3}$$

where t was the ground truth segmentation, p was the predicted multi-labeled segmentation, and i was the number of pixels in the image I. The DSC was computed separately for GTVp and GTVn, while the average value of these two DSCs was taken as the DSC for GTVp + GTVn. For the 3-fold and 5-fold CV, the averaged DSC scores for GTVp, GTVn and GTVp + GTVn are shown in Table 2.

In general, all experiments could achieve an average dice score for GTVp and GTVn around 0.65 (with an example of predicted segmentations in Fig. 2(A)). The 5-fold CV could achieve slightly better average scores than the 3-fold CV. The incorporation of multi-label cross-entropy loss yielded limited improvements for the average dice scores. We also noted the dice scores of GTVp were always higher than those of GTVn. In several cases, for example, CHUP-048 in Fig. 2(B), the GTVn was located in a region with comparably high SUV, leading to false positive predictions. In other cases, for example, CHUM-001 in Fig. 2(C), the low SUV region of the PET contained GTVn, leading to false negative predictions. Therefore, only PET/CT images might not be sufficient for GTVn auto-segmentation and some other clinical information would be required. For the testing dataset (359 patients), the segmentation performance was evaluated with another aggregated DSC [8] by the Hecktor organizer. As a result, our best model (the average prediction from 5-fold cross validation) achieved an average aggregated DSC for GTVp and GTVn of 0.70.

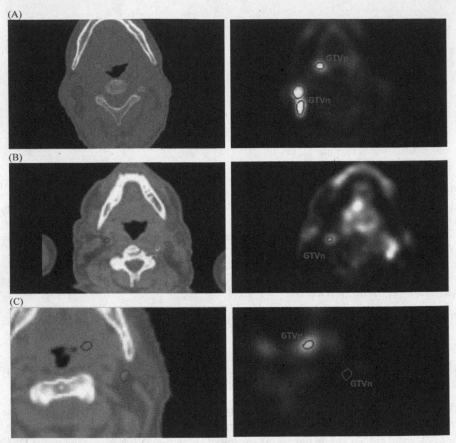

Fig. 2. Illustration of predicted segmentations for GTVp and GTVn (A), false positive (B) and false negative (C) predictions of GTVn. The true segmentation is contoured in purple lines on both PET and CT. The predicted GTVp is displayed in green and the predicted GTVn in red only on CT (Color figure online).

Table 2. Average (median) DSC scores from 3-fold and 5-fold cross-validation using two different loss functions for training

	Soft dice loss			Soft dice loss + cross entropy loss		
	GTVp + GTVn	GTVp	GTVn	GTVp + GTVn	GTVp	GTVn
3-fold	0.63 (0.67)	0.72(0.79)	0.57(0.69)	0.64 (0.71)	0.71(0.78)	0.58(0.68)
5-fold	0.66 (0.70)	0.71 (0.78)	0.58 (0.68)	0.66 (0.70)	0.72 (0.79)	0.57 (0.68)

3.3 Prognosis Task

The best HCIs for each model after 3-fold CV are summarized in Table 3. We observed that all models could produce CV HCIs with median values higher than 0.65 for RFS, while the PET-only model achieved the highest HCI value (0.70). In general, the PET-based models outperformed the CT-based models slightly. The CV HCIs of the joint PET/Seg model were lower than the ones obtained by the PET-only model, with a decline of median HCIs by 0.03. For the testing dataset (339 patients), our best model (the averaged predictions from the 3-fold cross validation PET-only models) achieved an HCI of 0.635.

Table 3. CV HCIs for different input schemes. The HCIs from the 3-fold CV and the average (median) HCI are reported.

Input	Fold 1	Fold 2	Fold 3	Average (median)
PET-only	0.73	0.66	0.71	0.70 (0.71)
PET/Seg	0.68	0.65	0.68	0.67 (0.68)
CT/Seg	0.65	0.65	0.63	0.64 (0.65)
PET/CT/Seg	0.74	0.63	0.67	1.68 (0.67)

4 Conclusion

In this study, the Retina Unet has been used to localize the HNC tumor (GTVp and GTVn) center from whole-body PET/CT scans, achieving a maximum difference of 4 cm and an average difference of (0.5, 0.8, 0.5) cm in (superior-inferior, lateral and anterior-posterior) direction. The PET/CTs were cropped based on their predicted tumor center and input into a multi-label Unet to segment GTVp and GTVn simultaneously. The proposed scheme produced encouraging performance and achieved an average GTVp/GTVn aggregated dice coefficient score of 0.70 on the testing dataset. Additionally, time-to-event analysis for RFS was implemented. In accordance with our previous study, the PET-only model achieved the best median CV HCI (0.69) over the other input combinations, indicating that PET appears to be a suitable modality for segmentation-free outcome prediction.

This study could be regarded as a step towards whole-body PET/CT tumor segmentation for HNC patients, but has some limitations. First, we did not conduct systematic analysis for the accuracy of tumor localization, which could be influenced by several factors, such as the box size of the retina Unet. Additionally, the proposed auto segmentation scheme was not end-to-end. The retina Unet was only applied to localize the tumor from whole body PET/CT scan. The potential of a fully automated end-to-end GTVp/GTVn segmentation with 2D and 3D Retina Unets and their comparison with the traditional Unets could be explored in future works.

Acknowledgements. This work was supported by the National Natural Science Foundation of China (NSFC) under Grant 61901087, China Postdoctoral Science Foundation under Grant 2019M663471, German Research Foundation (DFG) Research Training Group GRK 2274 'Advanced Medical Physics for Image-Guided Cancer Therapy', and Förderprogramm für Forschung und Lehre, Medical Faculty, LMU Munich, reg. no. 1084.

References

1. Economopoulou, P., Psyrri, A.: Head and neck cancers: essentials for clinicians, chap. 1 (2017)
2. Elkashty, O.A., Ashry, R., Tran, S.D.: Head and neck cancer management and cancer stem cells implication. Saudi Dent. J. **31**(4), 395–416 (2019)
3. Chajon, E., Lafond, C., Louvel, G., Castelli, J., Williaume, D., Henry, O., et al.: Salivary gland-sparing other than parotid-sparing in definitive head-and-neck intensity-modulated radiother-apy does not seem to jeopardize local control. Radiat. Oncol. (London, England). **8**, 132 (2013)
4. Baxi, S.S., Pinheiro, L.C., Patil, S.M., Pfister, D.G., Oeffinger, K.C., Elkin, E.B.: Causes of death in long-term survivors of head and neck cancer. Cancer **120**(10), 1507–1513 (2014)
5. Vallieres, M., Kay-Rivest, E., Perrin, L.J., Liem, X., Furstoss, C., Aerts, H., et al.: Radiomics strategies for risk assessment of tumour failure in head-and-neck cancer. Sci. Rep. 7(1), 10117 (2017)
6. Bogowicz, M., Riesterer, O., Stark, L.S., Studer, G., Unkelbach, J., Guckenberger, M., et al.: Comparison of PET and CT radiomics for prediction of local tumor control in head and neck squamous cell carcinoma. Acta Oncol. (Stockholm, Sweden). **56**(11), 1531–1536 (2017)
7. Han, M.W., Lee, H.J., Cho, K.J., Kim, J.S., Roh, J.L., Choi, S.H., et al.: Role of FDG-PET as a biological marker for predicting the hypoxic status of tongue cancer. Head Neck **34**(10), 1395–1402 (2012)
8. Andrearczyk, V., et al.: Overview of the HECKTOR Challenge at MICCAI 2021: Automatic Head and Neck Tumor Segmentation and Outcome Prediction in PET/CT Images. In: Andrea-rczyk, V., Oreiller, V., Hatt, M., Depeursinge, A. (eds.) Head and Neck Tumor Segmentation and Outcome Prediction. HECKTOR 2021. Lecture Notes in Computer Science, vol. 13209, pp. 1–37. Springer, Cham (2022). https://doi.org/10.1007/978-3-030-98253-9_1
9. Oreiller, V., Andrearczyk, V., Jreige, M., Boughdad, S., Elhalawani, H., Castelli, J., et al.: Head and neck tumor segmentation in PET/CT: the HECKTOR challenge. Med. Image Anal. **77**, 102336 (2022)
10. Jaeger, P.F., et al.: Retina U-Net: embarrassingly simple exploitation of segmentation super-vision for medical object detection. In: Machine Learning for Health Workshop, pp. 171–183. PMLR (2020)
11. Wang, Y., Lombardo, E., Avanzo, M., Zschaek, S., Weingartner, J., Holzgreve, A., et al.: Deep learning based time-to-event analysis with PET, CT and joint PET/CT for head and neck cancer prognosis. Comput. Methods Programs Biomed. **222**, 106948 (2022)
12. He, K., Zhang, X., Ren, S., Sun, J.: Deep residual learning for image recognition. In: Pro-ceedings of the IEEE Conference on Computer Vision and Pattern Recognition, pp. 770–778 (2016)
13. Isensee, F., Kickingereder, P., Wick, W., Bendszus, M., Maier-Hein, K.H.: Brain tumor seg-mentation and radiomics survival prediction: contribution to the brats 2017 challenge. In: Crimi, A., Bakas, S., Kuijf, H., Menze, B., Reyes, M. (eds.) BrainLes 2017. LNCS, vol. 10670, pp. 287–297. Springer, Cham (2018). https://doi.org/10.1007/978-3-319-75238-9_25
14. Kingma, D P., Ba, J.: Adam: a method for stochastic optimization. arXiv preprint arXiv:1412. 6980 (2014)
15. Ren, J., Huynh, B.N., Groendahl, A.R., Tomic, O., Futsaether, C.M., Korreman, S.S.: PET Normalizations to Improve Deep Learning Auto-Segmentation of Head and Neck Tumors in 3D PET/CT. In: Andrearczyk, V., Oreiller, V., Hatt, M., Depeursinge, A. (eds.) Head and Neck Tumor Segmentation and Outcome Prediction. HECKTOR 2021. Lecture Notes in Computer Science, vol 13209. Springer, Cham (2022). https://doi.org/10.1007/978-3-030-98253-9_7

16. Lin, T.-Y., Goyal, P., Girshick, R., He, K., Dollár PJae-p. Focal Loss for Dense Object Detection (2017) arXiv:1708.02002. https://ui.adsabs.harvard.edu/abs/2017arXiv170802002L
17. Girshick RJae-p. Fast R-CNN2015 (2015). arXiv:1504.08083. https://ui.adsabs.harvard.edu/abs/2015arXiv150408083G

HNT-AI: An Automatic Segmentation Framework for Head and Neck Primary Tumors and Lymph Nodes in FDG-PET/CT Images

Zohaib Salahuddin[1]([✉]), Yi Chen[1,2], Xian Zhong[1,3],
Nastaran Mohammadian Rad[1], Henry C. Woodruff[1,4], and Philippe Lambin[1,4]

[1] The D-Lab, Department of Precision Medicine, GROW-School for Oncology and
Reproduction, Maastricht University, Maastricht, The Netherlands
z.salahuddin@maastrichtuniversity.nl
[2] Key Laboratory of Intelligent Medical Image Analysis and Precise Diagnosis,
College of Computer Science and Technology, Guizhou University, Guiyang, China
[3] Department of Medical Ultrasonics, Institute of Diagnostic and Interventional
Ultrasound, The First Affiliated Hospital of Sun Yat-sen University,
Guangzhou, China
[4] Department of Radiology and Nuclear Medicine, GROW-School for Oncology,
Maastricht University Medical Center,
Maastricht, The Netherlands

Abstract. Head and neck cancer is one of the most prevalent cancers in the world. Automatic delineation of primary tumors and lymph nodes is important for cancer diagnosis and treatment. In this paper, we develop a deep learning-based model for automatic tumor segmentation, HNT-AI, using PET/CT images provided by the MICCAI 2022 Head and Neck Tumor (HECKTOR) segmentation Challenge. We investigate the effect of residual blocks, squeeze-and-excitation normalization, and grid-attention gates on the performance of 3D-UNET. We project the predicted masks on the z-axis and apply k-means clustering to reduce the number of false positive predictions. Our proposed HNT-AI segmentation framework achieves an aggregated dice score of 0.774 and 0.759 for primary tumors and lymph nodes, respectively, on the unseen external test set. Qualitative analysis of the predicted segmentation masks shows that the predicted segmentation mask tends to follow the high standardized uptake value (SUV) area on the PET scans more closely than the ground truth masks. The largest tumor volume, the largest lymph node volume, and the total number of lymph nodes derived from the segmentation proved to be potential biomarkers for recurrence-free survival with a C-index of 0.627 on the test set.

Team Name: The_DLab

Keywords: 3D UNet · Grid-attention · Residual networks · Squeeze-and-excitation · Segmentation biomarkers

Y. Chen and X. Zhong—These authors contributed equally as second authors.

1   Introduction

Head and Neck (H&N) cancers are among the most common cancers worldwide, including a group of tumors arising in the lip, oral cavity, pharynx, larynx, and paranasal sinuses [1]. The prognosis of H&N cancers varies greatly with 5-year survival ranging from 85.1% in patients with the localized disease to 40.1% in those with distant disease [2]. This variance in outcomes emphasizes the importance of accurate and timely diagnosis and staging, where imaging plays a crucial role. ^{18}F-FluoroDeoxyGlucose (FDG)-Positron Emission Tomography (PET) and Computed Tomography (CT) are two important imaging modalities for the initial staging and follow-up of H&N cancers. Nowadays, there is an increasing need for the development of automatic segmentation algorithms for H&N tumor diagnosis and staging, gross tumor volume (GTV) delineations in radiotherapy planning, as well as establishment and validation of radiomics models. PET and CT focused on metabolic and morphological characteristics, respectively, which may provide complementary and synergistic information for cancerous lesion segmentation as well as tumor characteristics predictive for patient outcome. The manual delineation of tumors in head and neck cancer suffers from inter-observer variability [3]. Hence, it is essential to develop automatic segmentation tools for H&N cancers that are accurate, fast, robust, and reproducible.

Deep neural networks are becoming popular due to the increasing amount of available data and computational resources. Convolutional Neural Networks (CNN) constitute a class of deep neural networks that have demonstrated state-of-the-art performance on a variety of medical image segmentation challenges, e.g., multi-center, multi-vendor, and multi-disease cardiac segmentation (M&Ms) challenge [4], multi-modality whole heart segmentation [5], and auto segmentation for thoracic radiation treatment planning [6]. No-new-UNet (nnUNet) has emerged as a self-configuring method that makes automatic design choices related to pre-processing, network architecture, and hyper-parameter tuning. It has demonstrated state-of-the-art performance in many medical image segmentation challenges [7].

The first and second editions of HEad and neCK TumOR (Hecktor) segmentation and outcome prediction challenge were held in 2020 and 2021 [8] [11]. These challenges aim to provide multi-centric data and use a standardized evaluation criteria to develop and validate automatic segmentation tools for H&N primary tumors in FDG-PET and CT images. The new edition of Hecktor 2022 extends Hecktor 2021 challenge by adding the additional task of lymph nodes segmentation (GTVn) along with the primary tumor (GTVp) segmentation, and it also provides data from three additional centers. In Hecktor 2021, a well-tuned 3D nn-UNet with squeeze-and-excitation normalization [13] demonstrated the best segmentation performance on the test set [12].

In this paper, we propose an HNT-AI segmentation framework based on nnUNet that incorporates residual blocks, squeeze-and-excitation channel-wise attention, and grid-attention gates. We also investigate if the primary tumor and lymph node segmentations produced by the proposed algorithm can serve as valuable biomarkers for recurrence-free survival prediction.

2   Material and Methods

2.1   Data

The training dataset for GTVp and GTVn segmentation comes from 7 different centers and consists of 524 training cases. A CT image and a corresponding registered FDG-PET image are provided for each case. The clinical information about each patient, including center, gender, age, tobacco and alcohol consumption, performance status (Zubrod), and HPV status is provided for 489 training cases. Additionally, the test dataset for evaluating the segmentation performance consists of 359 cases and it comes from 3 centers. Training cases from two of these centers are present in the training dataset.

2.2   Data Preprocessing

The CT images and the corresponding PET images are first resampled to the maximal bounding box covered by the field of view of both modalities. The CT images have a higher image resolution than the corresponding PET images. The median resolution of CT images in the training set was $0.98 \times 0.98 \times 3.27$ mm^3, and the mean resolution of PET images in the training set was $4.26 \times 4.26 \times 3.27$ mm^3. We resample both the CT and the PET images to the resolution of $1 \times 1 \times 3$ mm^3. The images are resampled with spline interpolation, and the corresponding segmentations are resampled using nearest-neighbor interpolation. We clip the intensity values of CT images at 0.5 and 99.5 percentiles. We apply z-score normalization on the CT images and min-max normalization on the PET images.

2.3   Network Architecture

The proposed architecture is shown in Fig. 1. This architecture is a modified adaption of the nnUNet architecture [7]. Each encoder block comprises of residual skip connections [14]. Each convolutional block consists of $3 \times 3 \times 3$ convolutional layer following by drop-out layer with p = 0.5, instance normalization and LeakyReLU activation function. Squeeze-and-excitation channel-wise attention mechanism is employed at each layer to learn important features by fusing spatial and channel-wise features [15]. Grid-attention gates enable the network to identify spatially important areas in the network and consequently aid in false positive reduction [16]. Grid attention gates are incorporated at each skip connection.

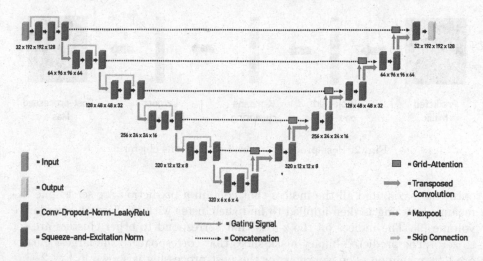

Fig. 1. The proposed architecture for 3D-UNet with skip connections, squeeze-and-excitation channel-wise attention mechanism and grid-attention gates.

2.4 Experimental Settings

We used a single RTX 3090 Graphics Card for training. The batch size was set at 2. Adam optimizer was used with an initial learning rate of 10^{-3}. Cosine annealing scheduler was used to reduce the learning rate from 10^{-3} to 10^{-6} every 30 epochs. The model was trained for 250 epochs. The training process took 9 h to complete. The patch size was set to be $192 \times 192 \times 128$. Data augmentation comprising of random rotation ($-15°$ to $+15°$), random scaling (0.85 to 1.15), elastic deformation, addition of Gaussian noise (0 to 0.1), mirroring, and gamma correction (0.75 to 1.25) was used to avoid overfitting. Five-fold cross-validation was used to find the best network architecture and for hyperparameter tuning.

2.5 Loss Function

The loss function comprised of multi-class dice loss [17] and cross entropy loss. The network is trained with deep supervision; thus, the computation of loss function occurs at each decoder block. This allows the training to occur at each layer. The loss computed at each decoder block is assigned a decreasing weight from higher resolution to lower resolution.

2.6 Post Processing

We observe false positive predictions for GTVp and GTVn outside the area of interest in the brain and the lung region. Furthermore, we observed small false positive predictions that occur at a distance from the correct ground truth prediction. To reduce the false positives, we projected the predicted binary mask

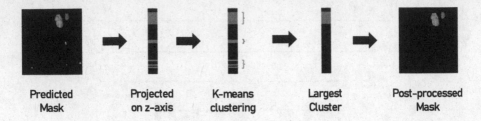

Fig. 2. Post-processing based on k-means clustering.

onto the z-axis, and all the indices that contain a prediction are set to one. K-means clustering is then applied to find the cluster with the largest cumulative volume [9]. The indices on the z-axis that correspond to other clusters are set to zero. The predicted binary mask slices that correspond to the largest cluster are left remaining. The workflow for the post-processing is shown in Fig. 2.

2.7 Evaluation Metric

Aggregate dice score (DSC_{agg}) is used as an evaluation metric. DSC_{agg} is defined as:

$$
DSC_{agg} = \frac{\sum\limits_{i}^{N}\sum\limits_{k} \hat{y_{i,k}} y_{i,k}}{\sum\limits_{i}^{N}\sum\limits_{k} (\hat{y_{i,k}} + y_{i,k})}
$$

where $y_{i,k}$ is the ground truth for GTV_p or GTV_n and $\hat{y_{i,k}}$ is the corresponding prediction. This metric is used because some images do not contain any tumor or lymph node. The intersections and unions for all the images are accumulated and then DSC_{agg} is calculated at the end.

3 Results

3.1 Five-Fold Cross-Validation

Table 1 shows the performance of modified 3D-UNet architecture for five-fold cross-validation. We started with nnUNet as a baseline. The proposed model that contains residual skip connections, squeeze-and-excitation channel-wise attention mechanism and grid attention gates achieves the best performances. Figure 3 shows the qualitative performance of the proposed algorithm on an image (CHUV-021) from the validation set. As shown, the predicted segmentation mask closely follows the boundary of the area with high SUV intake in the PET images compared to the ground truth label.

Table 1. Aggregated Dice DSC_{agg} for GTV_n and GT_n for five-fold cross validation.

Model	$DSC_{agg}\ GTV_p$	$DSC_{agg}\ GTV_n$	$Mean\ DSC_{agg}$
nnUNet	0.752 ± 0.471	0.731 ± 0.621	0.742 ± 0.546
3D UNet + Resnet	0.751 ± 0.508	0.744 ± 0.419	0.748 ± 0.463
3D UNet + SE	0.768 ± 0.398	0.759 ± 0.507	0.764 ± 0.452
Proposed	$\mathbf{0.777 \pm 0.351}$	$\mathbf{0.768 \pm 0.522}$	$\mathbf{0.773 \pm 0.437}$

Table 2. Aggregated Dice DSC_{agg} for GTV_n and GT_n of the three attempts on the test set

Model	$DSC_{agg}\ GTV_p$	$DSC_{agg}\ GTV_n$	$Mean\ DSC_{agg}$
nnUNet	0.756	0.729	0.743
3D UNET (Resnet + SE + Grid-Attention)	0.768	0.758	0.763
Ensemble	0.774	0.758	0.767

3.2 Test Set

Table 2 shows the result of our three attempts on the unseen test set. In the first attempt, we benchmarked the performance of nnUNet. For the second submission, 90% of the dataset was used for training and 10% of the dataset was used for validation. The proposed 3D UNet model with residual skip connections, squeeze-and-excitation channel-wise attention and grid attention gates was trained for 1000 epochs. For the final attempt, we combined the model used in attempt 2 with the five other models from five-fold cross-validation. The ensemble model was obtained by taking an average of the soft-max probabilities of the six models.

3.3 Tumor and Lymph Node Volume as a Biomarker for Prognosis

The prognostic value of tumor and lymph node volume derived from the segmentation are summarized in Table 3. Five-fold cross-validation shows that the model combining the largest tumor volume, largest lymph node volume and number of lymph nodes leads to the highest C-index of 0.597 ± 0.056 for recurrence-free survival (RFS) prediction. The model showed a C-index of 0.627 in the test set.

4 Discussion

The proposed 3D-UNet architecture based on residual skip connections, squeeze-and-excitation channel-wise attention, and grid-attention gates outperformed other network configurations during five-fold cross-validation. The proposed algorithm achieved $DSC_{agg}\ GTV_p$ score of 0.777 ± 0.351 and $DSC_{agg}\ GTV_n$ score of 0.768 ± 0.522 during five-fold cross-validation. On the unseen test set, the algorithm obtained $DSC_{agg}\ GTV_p$ score of 0.768 and $DSC_{agg}\ GTV_p$ score of

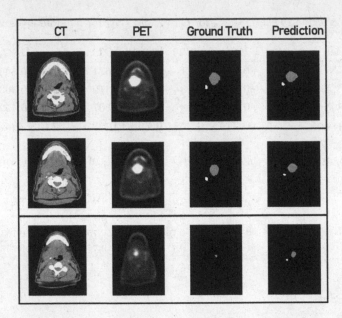

Fig. 3. Qualitative analysis of the predicted GTV_p and GTV_n predictions. The predicted mask tends to follow the areas of high SUV uptake more closely as compared to the ground truth labels.

Table 3. Prognostic value of tumor and lymph node volume for prediction of recurrence-free survival

Variables in the model	C–index $\pm$ SD
Five-fold Cross-Validation Results	
Largest Tumor Volume	0.565 ± 0.069
Largest Lymph Node Volume	0.534 ± 0.083
Number of Lymph Nodes	0.564 ± 0.033
Largest Tumor Volume + Largest Lymph Node Volume	0.570 ± 0.059
Largest Tumor Volume + Largest Lymph Node Volume + Number of Lymph Nodes	$\mathbf{0.597 \pm 0.056}$
Test Set Results	
Largest Tumor Volume + Largest Lymph Node Volume + Number of Lymph Nodes	0.627

0.758. Furthermore, the largest segmented tumor volume, the largest segmented lymph node volume and the total number of segmented lymph nodes appeared to be promising biomarkers for recurrence-free survival. These three features obtained a C-index of 0.627 on the external test set for RFS prediction. The proposed model demonstrates superior performance on the external test set as compared to five-fold cross-validation. This could not be investigated due to the absence of labels for the external test set.

This study has some limitations. It is important to determine that the developed algorithms are fair with respect to age, gender, and other biases. Therefore, metrics should also be calculated for these subgroups to report any biases. We observed qualitatively that in some of the cases, the proposed segmentation algorithm performs a better delineation. Therefore, an *in silico* trial that can obtain a preference score to determine whether the radiologists prefer automatic segmentation or manual delineations needs to be conducted [18]. We also observed that false positives occur in areas where there is no probability of the tumor occurring. The segmentation algorithm can be made more interpretable by including an anatomical prior that calculates the probability of tumor occurrence at each pixel value [19]. Furthermore, we also need to estimate uncertainty to avoid silent failures on out-of-distribution cases [20].

5   Conclusion

In this paper, we proposed a segmentation framework demonstrating promising performance for segmenting primary tumors and lymph nodes in head and neck cancer. We also validated the quality of the segmentation by using the largest segmented tumor and lymph node volumes as biomarkers for recurrence-free survival prediction. In the future, we need to increase interpretability by incorporating anatomical priors, estimating uncertainty, and calculating the fairness of the algorithm by evaluating the algorithm with respect to biases.

Acknowledgements. We acknowledge financial support from ERC advanced grant (ERC-ADG-2015 n° 694812 - Hypoximmuno), ERC-2020-PoC: 957565-AUTO. DISTINCT. Authors also acknowledge financial support from the European Union's Horizon 2020 Research and Innovation Programme under grant agreement: ImmunoSABR n° 733008, MSCA-ITN-PREDICT n° 766276, CHAIMELEON n° 952172, EuCanImage n° 952103, JTI-IMI2-2020-23-two-stage IMI-OPTIMA n° 101034347, and TRANSCAN Joint Transnational Call 2016 (JTC2016 CLEARLY n° UM 2017-8295). This work was supported by the Dutch Cancer Society (KWF Kankerbestrijding), Project number: 14449/2021-PoC. This study has received funding from the National Natural Science Foundation of China (No. 92059201). This study has received funding from the China Scholarship Council (No. 202006675008).

References

1. Pfister, D.G., et al.: Head and neck cancers, version 2.2020, NCCN clinical practice guidelines in oncology. J. Nat. Compr. Cancer Netw. **18**(7), 873–898 (2020)
2. Marcus, C., et al.: PET imaging for head and neck cancers. Radiol. Clin. **59**(5), 773–788 (2021)
3. van der Veen, J., Gulyban, A., Nuyts, S.: Interobserver variability in delineation of target volumes in head and neck cancer. Radiother. Oncol. **137**, 9–15 (2019)
4. Campello, V.M., et al.: Multi-centre, multi-vendor and multi-disease cardiac segmentation: the M&Ms challenge. IEEE Trans. Med. Imaging **40**(12), 3543–3554 (2021)

5. Zhuang, X., et al.: Evaluation of algorithms for multi-modality whole heart segmentation: an open-access grand challenge. Med. Image Anal. **58**, 101537 (2019)
6. Yang, J., et al.: Autosegmentation for thoracic radiation treatment planning: a grand challenge at AAPM 2017. Med. Phys. **45**(10), 4568–4581 (2018)
7. Isensee, F., et al.: nnU-Net: a self-configuring method for deep learning-based biomedical image segmentation. Nat. Methods **18**(2), 203–211 (2021)
8. Oreiller, V., et al.: Head and neck tumor segmentation in PET/CT: the HECKTOR challenge. Med. Image Anal. **77**, 102336 (2022)
9. Kodinariya, T.M., Makwana, P.R.: Review on determining number of Cluster in K-means clustering. Int. J. **1**(6), 90–95 (2013)
10. Andrearczyk, V., et al.: Overview of the HECKTOR challenge at MICCAI 2022: automatic head and neck tumor segmentation and outcome prediction in PET/CT. In: Head and Neck Tumor Segmentation and Outcome Prediction (2023)
11. Andrearczyk, V., et al.: Overview of the HECKTOR challenge at MICCAI 2021: automatic head and neck tumor segmentation and outcome prediction in PET/CT images. 3D Head and Neck Tumor Segmentation in PET/CT Challenge. Springer, Cham (2021)
12. Xie, J., Peng, Y.: The head and neck tumor segmentation based on 3D U-Net. In: Andrearczyk, V., Oreiller, V., Hatt, M., Depeursinge, A. (eds.) HECKTOR 2021. LNCS, vol. 13209, pp. 92–98. Springer, Cham (2022)
13. Iantsen, A., Visvikis, D., Hatt, M.: Squeeze-and-excitation normalization for automated delineation of head and neck primary tumors in combined PET and CT images. In: Andrearczyk, V., Oreiller, V., Depeursinge, A. (eds.) HECKTOR 2020. LNCS, vol. 12603, pp. 37–43. Springer, Cham (2021)
14. He, K., Zhang, X., Ren, S., Sun, J.: Identity mappings in deep residual networks. In: Leibe, B., Matas, J., Sebe, N., Welling, M. (eds.) ECCV 2016. LNCS, vol. 9908, pp. 630–645. Springer, Cham (2016). https://doi.org/10.1007/978-3-319-46493-0_38
15. Hu, J., Shen, L., Sun, G.: Squeeze-and-excitation networks. In: Proceedings of the IEEE Conference on Computer Vision and Pattern Recognition (2018)
16. Schlemper, J., et al.: Attention gated networks: learning to leverage salient regions in medical images. Med. Image Anal. **53**, 197–207 (2019)
17. Sudre, C.H., Li, W., Vercauteren, T., Ourselin, S., Jorge Cardoso, M.: Generalised dice overlap as a deep learning loss function for highly unbalanced segmentations. In: Cardoso, M.J., et al. (eds.) DLMIA/ML-CDS -2017. LNCS, vol. 10553, pp. 240–248. Springer, Cham (2017). https://doi.org/10.1007/978-3-319-67558-9_28
18. Primakov, S.P., et al.: Automated detection and segmentation of non-small cell lung cancer computed tomography images. Nat. Commun. **13**(1), 1–12 (2022)
19. Salahuddin, Z., et al.: Transparency of deep neural networks for medical image analysis: a review of interpretability methods. Comput. Biol. Med. **140**, 105111 (2022)
20. Mehrtash, A., et al.: Confidence calibration and predictive uncertainty estimation for deep medical image segmentation. IEEE Trans. Med. Imaging **39**(12), 3868–3878 (2020)

Head and Neck Tumor Segmentation with 3D UNet and Survival Prediction with Multiple Instance Neural Network

Jianan Chen[1,2] and Anne L. Martel[1,2(✉)]

[1] Department of Medical Biophysics, University of Toronto, Toronto, ON, Canada
`geoff.chen@mail.utoronto.ca`
[2] Physical Science Platform, Sunnybrook Research Institute, Toronto, ON, Canada
`anne.martel@sunnybrook.ca`

Abstract. Head and Neck Squamous Cell Carcinoma (HNSCC) is a group of malignancies arising in the squamous cells of the head and neck region. As a group, HNSCC accounts for around 4.5% of cancer incidences and deaths worldwide. Radiotherapy is part of the standard care for HNSCC cancers and accurate delineation of tumors is important for treatment quality. Imaging features of Computed Tomography (CT) and Positron Emission Tomography (PET) scans have been shown to be correlated with survival of HNSCC patients. In this paper we present our solutions to the segmentation task and recurrence-free survival prediction task of the HECKTOR 2022 challenge. We trained a 3D UNet model for the segmentation of primary tumors and lymph node metastases based on CT images. Three sets of models with different combinations of loss functions were ensembled to generate a more robust model. The softmax output of the ensembled model was fused with co-registered PET scans and post-processed to generate our submission to task 1 of the challenge, which achieved a 0.716 aggregated Dice score on the test data. Our segmentation model outputs were used to extract radiomic features of individual tumors on test data. Clinical variables and location of the tumors were also encoded and concatenated with radiomic features as additional inputs. We trained a multiple instance neural network to aggregate features of individual tumors into patient-level representations and predict recurrence-free survival rates of patients. Our method achieved an AUC of 0.619 for task 2 on the test data (Team name: SMIAL).

Keywords: Head and Neck · Segmentation · Outcome prediction · Radiomics · Multiple instance learning

1 Introduction

Head and Neck Squamous Cell Carcinoma (HNSCC) is a group of malignancies arising in the squamous cells of the head and neck region [7]. As a group, HNSCC accounts for around 4.5% of cancer incidences and deaths worldwide. Over half of HNSCC patients are diagnosed after the disease has spread to regional lymph nodes in the neck and nodal involvement is one of the most important prognostic factors in HNSCC. As a result, front-line therapy typically involves surgical removal of the primary tumor and extensive neck dissections followed by chemoradiotherapy. Sensitive detection of regional nodal involvement during initial workup and accurate delineation of primary and metastatic lesions is required for treatment planning. Patients who do not qualify for surgery may be treated with radiotherapy instead. Given the morbidity of treatment for HNSCC and the poor outcome of patients with recurrent disease, it is crucial to stratify patients upfront to potentially escalate or de-escalate treatment. While few targeted therapies of HNSCC are currently used in the clinic, predictive modeling could also aid in the selection of patients for trials of novel molecular therapies.

The HEad and neCK TumOR segmentation and outcome prediction in PET/CT images challenge (HECKTOR) has been hosted for three years to offer participants opportunities to develop and validate tumor segmentation and outcome prediction algorithms based on Head and Neck PET/CT data acquired from patients with oropharyngeal head and neck cancer who underwent radiotherapy treatment planning [1]. The 2022 edition includes a task (Task 1) for the segmentation of Head and Neck Gross Tumor Volumes (GTVp) and nodal Gross Tumor Volumes (GTVn) and another task (Task 2) for predicting recurrence-free survival (RFS) of patients.

In recent years, automatic tumor segmentation algorithms based on neural networks have shown great success in segmenting organs and tumors in PET/CT images. The UNet and its variants has been the most popular group of models [9,10]. Our solution for Task 1 involved an ensembling of three groups of nnUNet models trained on CT images with different loss functions followed by fusion with PET images and post-processing [6].

Quantitative imaging features have been shown to be correlated with survival of HNSCC patients [2,11]. We extracted radiomic features based on segmentations generated from Task 1 and concatenated them with clinical features and relative location of tumors as inputs. Since most patients in the dataset have more than one tumor, we trained an multiple instance neural network to take characteristics of all individual tumors into account [3,4].

2   Methods

2.1   Image Pre-processing

Following the winning solution in HECKTOR2021 for task 1 [13], we clipped the CT image intensities into [−200, 200] Hounsfield Units (HU) and normalized them using z-score normalization. CT and PET images were resampled to [1 mm, 1 mm, 1 mm] using third-order spline interpolation. PET image intensities were clipped into [0, 4] Standardized Uptake Values (SUV) to reduce the impact of pixels with high SUVs. PET images were resampled to the exact spacing and coordinate system of the corresponding CT images using nearest neighbour interpolation.

2.2   Segmentation Model

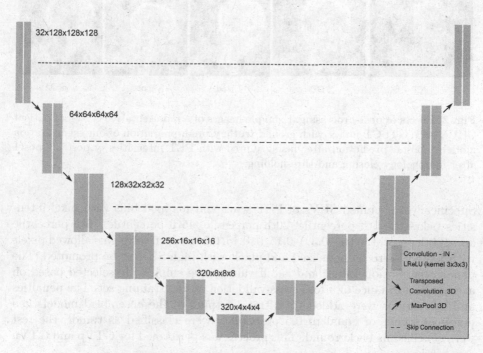

Fig. 1. Overview of the segmentation network structure. IN: instance normalization; LReLU: LeakyReLU activation.

For the segmentation model we used the default network structure of nnUNet (Fig. 1) [6], which contains an encoder and a decoder each composed of convolution layers, instance normalization layers, leaky ReLU activation layers and skip connections. The size of feature maps are displayed next to the corresponding encoder layers.

Various loss functions have been proposed to address different challenges in medical image segmentation and compound losses have been shown to be more robust [8]. In this task we trained 3 groups of models with Dice + Cross-entropy loss, Dice + Focal loss and Dice + TopK loss (k = 10th percentile), respectively. We ensembled 15 models (3 loss functions each with 5 fold cross-validations) by averaging their softmax predictions.

2.3   Image Post-processing

Predictions from the ensembled model were fused with pre-processed PET scans to generate corrected predictions, with the goal of mimicking the segmentation process of clinicians, i.e. ambiguous pixels in the CT were verified in the PET.

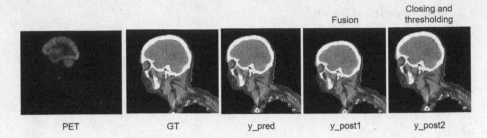

Fig. 2. Effects of post-processing. Example images of a representative GTVn of patient CHUM-001. GT: CT image with ground truth; y_pred: prediction of our segmentation model; y_post1: prediction after fusing y_pred with PET intensities; y_post2: y_post1 after morphology closing and thresholding.

Specifically, we binned SUVs of PET scans and mapped bins with pixel intensities below the 5th percentile, 20th percentile, 45th percentile, 70th percentile and 100th percentile to -0.3, -0.15, 0, 0.15, 0.3, respectively. This allowed pixels with low SUVs to be penalized and pixels with high SUV to be promoted. The penalty/promotion values and cutoff values were empirically selected based on average performance of 5-fold cross-validation on the training set. The penalties and promotions were added to softmax outputs of the ensembled model then pixels of higher or equal to 0.5 probability were classified as tumor, the rest were classified as background. This process was performed for GTVp and GTVn separately.

We observed that for some tumors, mostly lymph node metastases, the PET scan could display high SUV around the tumor boundary and low SUV in the tumor core, possibly due to necrosis. Fusing PET and CT in this situation led to false-negative predictions in tumor cores, therefore we performed a morphology closing operation with a ball of radius of 2 pixels as the structuring element. Finally, thresholding with HU larger than -50 was carried out as the last processing step to exclude false positives, based on the idea that cancerous tissues should have intensities higher than -50 HU. The effect of post-processing is demonstrated in Fig. 2.

2.4   Feature Extraction

We separated the predictions of our segmentation model into multiple instances, each representing one lesion. We then extracted radiomic features from test CT and test PET images for each individual lesion. 103 original (i.e. without filtering) imaging features, including shape, intensity, grey level co-occurrence matrix, grey level run-length matrix, grey level size zone matrix, grey level dependence matrix and neighbourhood grey tone difference matrix features were extracted. Tumor location in the image, represented by the center of mass coordinates carried tumor spatial information and was an indicator of whether the tumor was primary or metastatic. Therefore, we encoded the relative locations (calculated with pyradiomics) of tumors as location features. Although we didn't register the head and neck regions to an atlas, we observed improvements by introducing these location features. We excluded clinical features with incomplete records so that three features: age, sex and weight are incorporated in the feature sets, resulting in a total of 212 features for each lesion.

2.5   Prediction Model

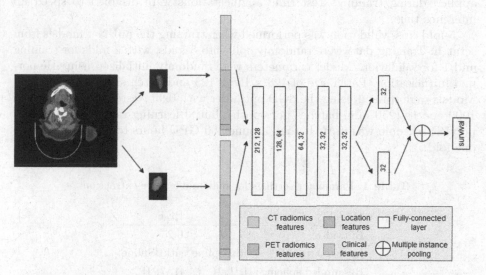

Fig. 3. Recurrence-free survival prediction based on multiple instance learning.

Since most patients in the dataset had more than 1 lesion (counting both primary and metastatic lesions), we approached the RFS prediction problem with a method specifically designed for multifocal/metastatic cancer outcome prediction [3,4]. We used a multiple instance learning setting where a patient was treated as a bag and lesions were treated as instances. Features of each lesion were passed into the network to generate an instance-level representation and all the instance representations from the same patient were aggregated to generate the patient-level representation for predicting patient risk. We followed the network design in previous work [4] but modified the number of layers and hidden units as more cases and more features were available (Fig. 3). Average pooling was chosen for aggregating instance-level representations as it achieved the best performance in cross validations when compared to max pooling and attention-based pooling [5]. Cross entropy loss was used to train the multiple instance neural network.

3   Experimental Results

3.1   Implementation Details

Segmentation. The experiments were implemented in a Linux environment with a NVIDIA V100 Tensor Core GPU with 16 GB VRAM. We used python 3.7, pytorch 1.11.0, torchvision 0.11.1 and nnunet 1.7.0 for the segmentation model. Augmentations, including rotation in $[-30, 30]$, scaling with $[0.7, 1.4]$, mirroring, Gamma correction of $[0.7, 1.3]$ and Gaussian noise of $[0, 0.15]$, were applied during training. Test time augmentations were disabled to speed up inference time.

5-fold cross validation was performed when training the nnUNet models from scratch. Training data were randomly split into 5 folds, with 4 folds for training and 1 for validation. Model parameters were randomly initialized using He normal initialization. Patch size of $128 \times 128 \times 128$ and batch size of 2 were used. Models were trained using the SGD optimizer with 0.99 Nesterov momentum in 1000 epochs (250 mini-batches per epoch). Initial learning rate was 0.01 and a polyLR schedule was used. It took around 100 GPU hours to train a model for one fold.

Table 1. Parameter settings for radiomic feature extraction

Modality	CT	PET
Binwidth	15	0.25
Interpolator	sitkBSpline	sitkBSpline
Resampled spacing	[1, 1, 1]	[1, 1, 1]
Resegment range	[−3, 3]	[−3, 3]
Resegment mode	sigma	sigma
Voxel array shift	200	0

Survival Prediction. For RFS prediction we used python 3.7, pyradiomics 3.0.1[1] [12], SimpleITK 2.0.2, numpy 1.18.5 and pytorch 1.10.1. Parameters for feature extraction are summarized in Table 1.

5-fold cross validation was performed for training the multiple instance neural network models from scratch. Model parameters were randomly initialized using He normal initialization. A batch size of one was used to accommodate the multiple instance learning setting. Models were trained using the SGD optimizer with weight decay of 0.001 in 100 epochs. A fixed learning rate of 0.0003 was used. Patient survival was right-censored at 1460 days (4 years). It took around 3 min to train a model on one fold.

---

[1] https://pyradiomics.readthedocs.io/en/latest/index.html.

Table 2. Results of 5-fold cross validation in terms of Dice, Precision, Recall of GTVp and GTVn before and after post-processing. post1: fusing predictions with PET intensities; post2: post1 after morphology closing and thresholding.

	Dice_GTVp	Dice_GTVn	Precision_GTVp	Precision_GTVn	Recall_GTVp	Recall_GTVn
Fold1	0.563	0.615	0.608	**0.647**	0.628	**0.730**
Fold2	**0.614**	**0.620**	0.629	0.627	**0.674**	0.728
Fold3	0.597	0.590	0.628	0.624	0.666	0.689
Fold4	0.611	0.543	**0.673**	0.543	0.639	0.673
Fold5	0.604	0.607	0.640	0.631	0.659	0.720
Average	0.598	0.595	**0.636**	**0.653**	0.614	**0.708**
Average_post1	0.610	0.601	0.627	0.639	0.688	0.699
Average_post2	**0.612**	**0.603**	0.624	0.638	**0.696**	0.705

Table 3. Task 1 results on the test data for our two submissions.

	Dice_agg_GTVp	Dice_agg_GTVn	Dice_agg_average
Submission 1 (average ensemble)	0.676	**0.753**	0.714
Submission 2 (average ensemble with postprocessing)	**0.681**	0.751	**0.716**

3.2 Primary Tumor and Lymph Node Metastasis Segmentation

The performance of our segmentation model on the validation set and test set is summarized in Table 2 and Table 3, respectively. The average dice score in cross-validations is around 0.6 for both GTVp and GTVn, with relatively large variations across different folds. Adding the post-processing steps, namely post 1 (fusion with PET scan) and post 2 (post 1 with additional morphology closing and thresholding) improved the Dice score considerably for both GTVp and GTVn. On the held-out test set, our model achieved 0.71419 (ensembled model w/o post-processing) and 0.71591 (ensembled model w/ post-processing 2) in two runs, with much better dice score for GTVn compared to GTVp.

3.3 Recurrence-Free Survival Prediction

Table 4. Task 2 results on the test data for our three submissions.

	AUC
Submission 1 (ensembling over 5 folds with concordant output)	0.402
Submission 2 (ensembling over 5 folds with underfit models)	0.498
Submission 3 (model trained with the whole training set)	**0.619**

The performance of our recurrence-free survival (RFS) prediction model on test set is summarized in Table 4. We made mistakes in our first and second submissions. In our first submission, we trained a multiple instance neural network

(MINN) with 5-fold cross validation and ensembled the models from the 5 folds to make predictions on the test data. This approach, however, produced concordant predictions instead of anti-concordant predictions requested by the challenge, therefore getting one minus the actual AUC. The actual AUC for submission 1 should be 0.59820. In our second submission, we accidentally submitted results from underfit models, therefore getting an AUC close to 0.5. In our third submission, we took a more conservative approach and trained the model on the whole training set without cross-validation or model selection, getting a final AUC of 0.61877.

4   Discussion and Conclusion

In this work we trained an nnUNet model based on CT images to segment GTVp and GTVn for HNSCC. Although we found a large gap between aggregated Dice score of GTVp and GTVn, the model worked well on the test data. The post-processing we proposed improved the performance on the validation set by about 2%, and provided slight improvements on the test data despite the domain shift. Better performance might be achieved if the model was trained based on both modalities, but due to limited time and resources, we trained the model with CT data only.

For the outcome prediction challenge we trained a multiple instance neural network based on radiomic features and clinical features. We used segmentations from task 1. Using these inaccurate segmentation masks may have had a big impact on the performance of our outcome prediction model because it may have drastically changed the number of instances and instance features but despite this, our model still had prognostic value. If ground truth segmentation masks for the test data are released in the future, it would be interesting to assess the performance of the outcome model independently of errors in segmentation.

Acknowledgements. We acknowledge support of the Natural Sciences and Engineering Research Council of Canada (NSERC) and the Digital Research Alliance of Canada.

References

1. Andrearczyk, V., et al.: Overview of the HECKTOR challenge at MICCAI 2022: automatic head and neck tumor segmentation and outcome prediction in PET/CT images. In: Andrearczyk, V., et al. (eds.) HECKTOR 2022. LNCS, vol. 13626, pp. 1–30. Springer, Cham (2023)
2. Bogowicz, M., et al.: Comparison of PET and CT radiomics for prediction of local tumor control in head and neck squamous cell carcinoma. Acta Oncol. **56**(11), 1531–1536 (2017)
3. Chen, J., Cheung, H.M.C., Milot, L., Martel, A.L.: AMINN: autoencoder-based multiple instance neural network improves outcome prediction in multifocal liver metastases. In: de Bruijne, M., et al. (eds.) MICCAI 2021. LNCS, vol. 12905, pp. 752–761. Springer, Cham (2021). https://doi.org/10.1007/978-3-030-87240-3_72

4. Chen, J., Martel, A.L.: Metastatic cancer outcome prediction with injective multiple instance pooling. arXiv preprint arXiv:2203.04964 (2022)

5. Ilse, M., Tomczak, J., Welling, M.: Attention-based deep multiple instance learning. In: International Conference on Machine Learning, pp. 2127–2136. PMLR (2018)

6. Isensee, F., Jaeger, P.F., Kohl, S.A., Petersen, J., Maier-Hein, K.H.: nnU-Net: a self-configuring method for deep learning-based biomedical image segmentation. Nat. Methods 18(2), 203–211 (2021)

7. Johnson, D.E., Burtness, B., Leemans, C.R., Lui, V.W.Y., Bauman, J.E., Grandis, J.R.: Head and neck squamous cell carcinoma. Nat. Rev. Dis. Primers. 6(1), 1–22 (2020)

8. Ma, J., et al.: Loss odyssey in medical image segmentation. Med. Image Anal. 71, 102035 (2021)

9. Milletari, F., Navab, N., Ahmadi, S.A.: V-Net: fully convolutional neural networks for volumetric medical image segmentation. In: 2016 Fourth International Conference on 3D Vision (3DV), pp. 565–571. IEEE (2016)

10. Ronneberger, O., Fischer, P., Brox, T.: U-Net: convolutional networks for biomedical image segmentation. In: Navab, N., Hornegger, J., Wells, W.M., Frangi, A.F. (eds.) MICCAI 2015. LNCS, vol. 9351, pp. 234–241. Springer, Cham (2015). https://doi.org/10.1007/978-3-319-24574-4_28

11. Vallieres, M., et al.: Radiomics strategies for risk assessment of tumour failure in head-and-neck cancer. Sci. Rep. 7(1), 1–14 (2017)

12. Van Griethuysen, J.J., et al.: Computational radiomics system to decode the radiographic phenotype. Can. Res. 77(21), e104–e107 (2017)

13. Xie, J., Peng, Y.: The head and neck tumor segmentation based on 3D U-Net. In: Andrearczyk, V., Oreiller, V., Hatt, M., Depeursinge, A. (eds.) HECKTOR 2021. LNCS, vol. 13209, pp. 92–98. Springer, Cham (2021). https://doi.org/10.1007/978-3-030-98253-9_8

Deep Learning and Machine Learning Techniques for Automated PET/CT Segmentation and Survival Prediction in Head and Neck Cancer

Mohammad R. Salmanpour[1,2,3]([✉]), Ghasem Hajianfar[3,4], Mahdi Hosseinzadeh[3,5], Seyed Masoud Rezaeijo[6], Mohammad Mehdi Hosseini[7], Ehsanhosein Kalatehjari[8], Ali Harimi[8], and Arman Rahmim[1,2]

[1] Department of Physics and Astronomy, University of British Columbia, Vancouver, BC, Canada
msalman@bccrc.ca
[2] Department of Integrative Oncology, BC Cancer Research Institute, Vancouver, BC, Canada
[3] Technological Virtual Collaboration (TECVICO Corp), Vancouver, BC, Canada
[4] Rajaie Cardiovascular Medical and Research Center, Iran University of Medical Science, Tehran, Iran
[5] Department of Electrical and Computer Engineering, University of Tarbiat Modares, Tehran, Iran
[6] Department of Medical Physics, Faculty of Medicine, Ahvaz Jundishapur University of Medical Sciences, Ahvaz, Iran
[7] Department of Computer Engineering, Islamic Azad University, Shahrood Branch, Shahrood, Iran
[8] Department of Electrical Engineering, Islamic Azad University, Shahrood Branch, Shahrood, Iran

Abstract. Background: Accurate prognostic stratification as well as segmentation of Head-and-Neck Squamous-Cell-Carcinoma (HNSCC) patients can be an important clinical reference when designing therapeutic strategies. We set to enable automated segmentation of tumors and prediction of recurrence-free survival (RFS) using advanced deep learning techniques and Hybrid Machine Learning Systems (HMLSs).

Method: In this work, 883 subjects were extracted from HECKTOR-Challenge: ~60% of the total subjects were considered for the training and validation procedure, and the remaining subjects for external testing were employed to validate our models. PET images were registered to CT. First, a weighted fusion technique was employed to combine PET and CT information. We also employed Cascade-Net to enable automated segmentation of HNSCC tumors. Moreover, we extracted deep learning features (DF) via a 3D auto-encoder algorithm from PET and the fused image. Subsequently, we employed an HMLS including a feature selection algorithm such as Mutual Information (MI) linked with a survival prediction algorithm such as Random Survival Forest (RSF) optimized by 5-fold cross-validation and grid search. The dataset with DFs was normalized by the z-score technique. Moreover, dice score and c-Index were reported to evaluate the segmentation and prediction models, respectively.

© The Author(s), under exclusive license to Springer Nature Switzerland AG 2023
V. Andrearczyk et al. (Eds.): HECKTOR 2022, LNCS 13626, pp. 230–239, 2023.
https://doi.org/10.1007/978-3-031-27420-6_23

Result: For segmentation, the weighted fusion technique followed by the Cascade-Net segmentation algorithm resulted in a validation dice score of 72%. External testing of 71% confirmed our findings. DFs extracted from sole PET and MI followed by RSF enabled us to receive a validation c-index of 66% for RFS prediction. The external testing of 59% confirmed our finding.

Conclusion: We demonstrated that using the fusion technique followed by an appropriate automated segmentation technique provides good performance. Moreover, employing DFs extracted from sole PET and HMLS, including MI linked with RSF, enables us to perform the appropriate survival prediction. We also showed imaging information extracted from PET outperformed the usage of the fused images in the prediction of RFS.

Keywords: Survival prediction · Automated segmentation · Head and neck cancer · Fusion technique · Hybrid Machine Learning System

1 Introduction

Head and neck squamous cell carcinomas (HNSCCs) are the most common malignancies that arise in the head and neck, involving over 655000 people yearly worldwide. It is reported that half of these cases result in death [11]. Brain tumors diagnosed in the early stages play a critical role in successful treatment. Positron Emission Tomography (PET), Single-Photon Emission Computed Tomography (SPECT), Computed Tomography (CT), Magnetic Resonance Spectroscopy (MRS), and Magnetic Resonance Imaging (MRI) are common medical imaging procedures to provide accurate delineation of tumor volume, location, and metabolism [12]. Multimodality image techniques such as PET/CT provide more sensitivity and specificity than MRI or CT alone in head and neck (H&N) cancer imaging [13–16]. PET is used primarily to assess physiology, while CT is mainly used to assess anatomy.

Since different image modalities such as MRI, ultrasound, and PET include specific information (perspectives) of the same object, image fusion techniques enable us to combine two or more images to enhance the information content, improve the performance of object recognition systems by integrating many sources of imaging systems, help in sharpening the images, improve geometric corrections, enhance certain features that are not visible in either of the images, replace the defective data, complement the data sets for better decision making, and keep the integrity of important features or remove the noise [17]. It has been shown that fusing different images reduces ambiguity' and enhances the reliability of defect detection in both visual and qualitative evaluation [18]. As such, many studies [19]indicated that fusion techniques are considered a vital pre-processing phase for several applications such as outcome prediction, early diagnosis, segmentation (or delineation), and others.

Although handcrafted radiomics software (e.g., standardized software [20]) enables RF extraction from the regions of interest, utilizing deep learning (DL) algorithms; e.g. autoencoders, enable comprehensive extraction of deep imaging features (DF) from images [21]. The encoder layers extract important representations of the image using the feature learning capabilities of the convolutional neural networks. As such, application to different tasks using DFs extracted from DL algorithms has the potential to outperform other imaging features such as radiomics features (RF), which we investigate in the present work. In fact, autoencoders, as classification, captioning, and unsupervised learning algorithms, have been explored in the past to use such informative DFs for different predictions or clustering [22].

The HEad and neCK TumOR segmentation and outcome prediction from PET/CT images (HECKTOR) challenge aimed at identifying the best methods to leverage the rich bi-modal information in the context of H&N primary tumor segmentation and outcome prediction [7]. Manual tumor delineation methods are costly, tedious, and error-prone. An automated segmentation method for H&N tumors would greatly assist in optimizing personalized patient treatment plans. Artificial intelligence (AI) techniques, especially CL-based methods, are widely used to help accelerate the segmentation process [23]. The HECKTOR 2022 challenge allows participants to develop automated H&K tumor segmentation algorithms in PET/CT scans [7]. In the past few years, DL techniques based on convolutional neural networks (CNN) delivered excellent medical image analysis tasks, including image segmentation. Wang et al. [24] used 3D U-Net as the backbone architecture, on which a residual network is added to better capture image detail information. Xie et al. [25] adopted the 3D U-Net network to carry out automated segmentation of H&N tumors based on the dual modality PET-CT images. In a study [2], 3DUNETR (Unet with Transformers) and 3D-UNet have been utilized to segment HNSCC tumors automatically. In our previous studies [1], multiple fusion techniques were utilized to combine PET and CT information. Moreover, we employed 3D-UNet architecture and SegResNet (segmentation using autoencoder regularization) to improve segmentation performance.

Prediction of recurrence-free survival (RFS) [1, 2, 26], provides important prognostic information required for treatment planning. Recent radiomics studies based on PET/CT or fusion of images have been investigated to predict the survival outcomes in patients with H&N cancer. Although many studies [26, 27] have focused on the prediction of survival outcomes and stages through RFs in H&N cancer, this study is investigating DFs to enhance the prediction performance. Fusion DFs, as an emerging area of investigation, have up to now meant the exploration of fusing images in different ways and selecting the optimal one [28].

In this study, as elaborated next, we first registered PET to CT, and both images were then normalized. We also employed weighted fusion to combine PET and CT information. Moreover, we utilized a Cascade-Net to enable automated segmentation of HNSCC tumors. Subsequently, we utilized a DL algorithm to extract DFs from the fused image. Next, a Hybrid Machine Learning System (HMLS) including Mutual Information (MI) linked with a Random Survival Forest (RSF) was employed to predict RFS. Overall, we aimed to understand if and how the use of DFs can add value relative to the use of conventional hand-crafted RFs.

2   Material and Methods

2.1   Patient Data

883 subjects (total subjects) were extracted from 2022 HECKTOR Challenge. We employed ~60% of the total subjects for the training procedure, and the remaining subjects were employed to validate our models. In the pre-processing step, PET images were first resampled (registered) to CT, and then both images were normalized. Next, we employed a weighted fusion technique to combine CT and PET information. Figure 1 demonstrates examples of CT, PET, and the fused image generated via weighted images.

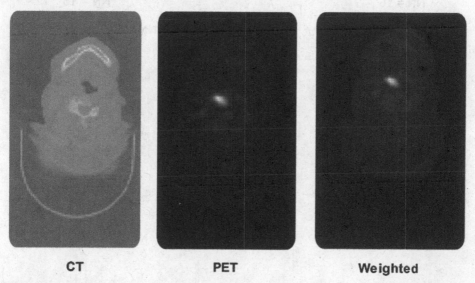

Fig. 1. Fusion process. Some examples of 1) CT image alone, 2) PET image alone, 3) fusion of PET and CT generated via weighted fusion technique

2.2   Feature Extraction Using Autoencoder Algorithm

In the DFs extraction framework, a 3D autoencoder neural network architecture [22] was used to extract DFs thoroughly. Generally, every autoencoder mainly comprises an encoder network and a decoder network. The encoding layer maps the input images to a latent representation or bottleneck, and the decoding layer maps this representation to the original images. So, the number of neurons in the input and output layers must be the same in an autoencoder. Also, the training label is the same as the input data.

The encoder follows typical convolutional network architecture, as shown in Fig. 2. It consists of three 3×3 convolutional layers, each followed by a leaky rectified linear unit (LeakyReLU) and a 2×2 max-pooling operation. The pooling layers are used to reduce the dimensions of the parameters and scale down the amount of computational performed in the network. The decoder path consists of three 3×3 convolutional layers, followed by a LeakyReLU and an up-sampling operation. We used a common loss function for the proposed autoencoder called binary cross-entropy. Thus, the proposed autoencoder was trained with a gradient-based optimization algorithm, namely Adam, to minimize the loss function. In the pre-processing step, PET images were first resampled to CT, and then both images were normalized. After fusing the preprocessed images, the 3D autoencoder model is employed to extract 5120 DFs via the bottleneck layer from the fused image and sole PET.

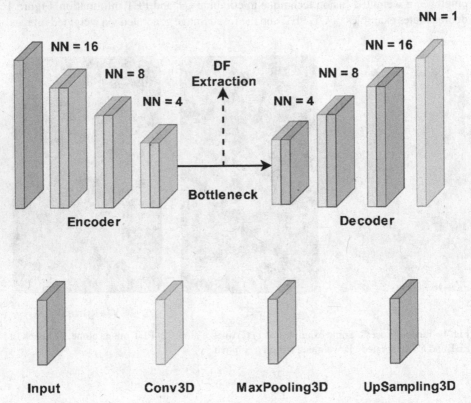

Fig. 2. Structure of our autoencoder model. It includes three 3 × 3 convolutional layers, each followed by a leaky rectified linear unit (LeakyReLU) and a 2 × 2 max-pooling operation. The decoder path includes three 3 × 3 convolutional layers, followed by a LeakyReLU and an up-sampling operation. NN: Number of neurons.

2.3   AI Algorithms

2.3.1   Cascade-Net Algorithm

In the Cascade-Net algorithm was proposed an attention mechanism for 3D medical image segmentation. This method, named segmentation-by-detection, is a cascade of a detection module linked with a segmentation module [29].

2.3.2 Mutual Information (MI) Algorithm

Hence the usage of feature selection algorithms enable improves prediction performance [4–6], in this study, we employed MI prior to the RSF algorithm to select the most relevant DFs. MI employed Somers' Dxy index to calculate the correlation between survival data and DFs [30].

2.3.3 Random Survival Forest (RSF) Algorithm

RSF employed Breiman's random forests method to predict the outcome. In random forests, randomization is introduced in two forms. First, a randomly drawn bootstrap sample of the data is employed to grow a tree. Second, at each node of the tree, a randomly selected subset of features (covariates) is selected as candidate features for splitting. Averaging over trees, in combination with the randomization used in growing a tree, enables random forests to approximate rich classes of functions while maintaining low generalization error [31]. We used RSF with hyperparameters 1200 HP; ntree: 100, 500, 1000; mtry: 1–10; nodesize: 1:20; splitrule: logrank, logrankscore.

2.4 Analysis Procedure

In preprocessing step, 883 PET images were first resampled to their CT images, and then both images were normalized. Subsequently, a weighted fusion technique was employed to combine the preprocessed PET and CT images. Next, a 3D autoencoder model was utilized to extract 5024 DFs through the bottleneck layer from the fused images and sole PET images, as elaborated in Sect. 2.2. In the first step of our analysis, we employed an automated DL segmentation algorithm such as Cascade-Net to segment HNSCC tumors, as explained in Sect. 2.3.1. In the second step of our analysis, we subsequently employed an HMLS, including a feature selection algorithm such as MI linked with a survival prediction algorithm such as RSF, optimized by 5-fold cross-validation and grid search, as elaborated in Sects. 2.3.2 and 2.3.3. After extracting DL features through autoencoder algorithms from the fused usage and sole PET, we normalized the dataset by the z-score technique. Moreover, an average dice score (average of H&N primary tumors (GTVp) and nodal Gross Tumor Volumes (GTVn)) and c-Index were reported to evaluate the segmentation and prediction models, respectively. Furthermore, we used ~80% of the total subjects to select the optimal model in a five-fold cross-validation process. Meanwhile, the remaining subjects were utilized for external testing of selected models. All codes were performed in Python 3.9. Moreover, MI and RSF were performed in R version 4.0 and mlr package version 2.18.

3 Results and Discussion

After the image preprocessing step, we first applied the images to a weighted fusion algorithm linked with a Cascade-Net segmentation algorithm. Thus, an averaged dice score of 72% ± 8% was provided in five-fold cross-validation. The external testing of 71% confirmed our finding. This study demonstrated enhanced task achievement for the automated segmentation of tumors in task 1, compared to our previous studies [1, 2] At the same time, in our past studies, we achieved dice scores around 0.63 and 0.68 through LP-SR Mixture fusion technique (the mixture of Laplacian Pyramid (LP) and Sparse Representation (SR) fusion techniques) and LP fusion technique linked with a 3D-UNet segmentation algorithm. In short, the usage of Cascade-Net segmentation algorithm, instead of 3D-UNet-based segmentation led to an improvement in the segmentation of H&N tumors.

Employing machine learning algorithms enables an enhancement in task prediction without explicit hand-coding [3]. Machine learning approaches aim to enable predictions by extracting statistically robust structures in the analyzed data. Employing datasets with high dimensions restricts us from utilizing most prediction algorithms, and it is important to reduce the dimension of the dataset to avoid overfitting. There are two known ways to reduce the dimension via i) attribute extraction algorithms and ii) feature selection algorithms. Attribute extraction algorithms aim to extract some relevant attributes from the original data so that the attributes are not similar to the features in the original dataset.

At the same time, feature selection algorithms enable subset selection of relevant features of the original dataset [4–6]. In this study, we employed FSA to select the relevant DFs. In the survival prediction task, the usage of the MI algorithm enabled us to select 100 relevant DFs. As such, RSF enables us to achieve a five-fold cross-validation c-Indexes of 0.66 ± 0.03 and 0.61 ± 0.04 through 100 selected DFs extracted from PET and the fused images. External testing c-Index of 0.59 confirmed our findings. At the same time, in our past effort [1] we achieved c-Indexes around 0.66 by employing LP-SR Mixture linked with GlmBoost (Gradient Boosting with Component-wise Linear Models). Moreover, in another study of ours [2] a c-Index of 0.68 and was obtained via an ensemble voting technique. The usage of DFs, instead of RFs, added a significant value to the prediction of survival, compared to our previous studies. As a result, DFs extracted from sole PET outperformed the DFs extracted from the fused images.

Furthermore, automated segmentation has the potential to allow radiation oncologists or physicians to improve treatment planning efficiency by reducing the time needed for tumor delineation as well as improving inter-observer reproducibility [7]. The ability to predict the period of the disease and the impact of interventions is essential to effective medical practice and healthcare management. Hence, accurately predicting a disease diagnosis can help physicians make more informed clinical decisions on treatment approaches in clinical practice [8]. The prediction of outcomes using quantitative image biomarkers from medical images (i.e. handcrafted RFs and DFs) has shown tremendous potential to personalize patient care in the context of H&N tumors [9]. The metabolic and morphological characteristics of H&N tumor tissue such as CT and PET images provide complementary and synergistic data for patient outcome prediction [10].

As a limitation of this study, future studies with a large sample size are suggested. Our study considered automated segmentation and survival prediction for two-images CT and PET; hence the proposed approaches can also be used for other related tasks in medical image analysis such as mpMR images, including T2 weighted image (T2W), diffusion-weighted magnetic resonance imaging (DWI), apparent diffusion coefficient (ADC), and dynamic contrast-enhanced magnetic resonance imaging (DCE-MRI). Moreover, this study can further benefit from more fusion techniques to combine PET and CT images. In addition, future studies should seek to employ more DL segmentation algorithms and further optimize these methods for improved H&N tumor segmentation performance in forthcoming iterations of the HECKTOR Challenge. Moreover, the usage of more FSAs or AEAs can be considered to construct new HMLSs. Furthermore, future studies can seek to employ more HMLSs to enhance the survival outcome prediction performance in forthcoming iterations of the HECKTOR Challenge. The novelty of this study is the usage of a fusion technique to generate new images from CT and PET images. Further, employing DL algorithms to generate DFs added significant value to the prediction of the survival outcome, compared to our previous studies.

4 Conclusions

This study presented the development and validation of DL models using a 3D Residual UNET architecture to segment H&N tumors in an end-to-end automated workflow based on PET/CT images and the fused image. Using a combination of pre-processing steps, and the DL algorithm, we achieve a dice score of ~72% for our best model. Our method notably improves upon the previous iteration of our model submitted in the 2021 HECKTOR Challenge. Moreover, this study investigated the prediction of survival via DFs and machine learning algorithms such as MI linked with an RSF, resulting in a c-Index of 0.66. Overall, we showed that the usage of image fusion followed by appropriate automated segmentation provides good performance. Moreover, employing HMLS, including the MI linked with RSF, enables us to perform the appropriate survival prediction.

Acknowledgment. This study was supported by the Technological Virtual Collaboration Corporation (TECVICO Corp.), Vancouver, BC, Canada, as well as the Natural Sciences and Engineering Research Council of Canada (NSERC) Discovery Grant RGPIN-2019–06467.

Code Availability. All codes are publicly shared at: https://github.com/Tecvico.

Conflict of Interest. The authors have no relevant conflicts of interest to disclose.

References

1. Fatan, M., Hosseinzadeh, M., Askari, D., Sheikhi, H., Rezaeijo, S.M., Salmanpour, M.R.: Fusion-based head and neck tumor segmentation and survival prediction using robust deep learning techniques and advanced hybrid machine learning systems. In: Andrearczyk, V., Oreiller, V., Hatt, M., Depeursinge, A. (eds.) HECKTOR 2021. LNCS, vol. 13209, pp. 211–223. Springer, Cham (2022). https://doi.org/10.1007/978-3-030-98253-9_20
2. Salmanpour, M.R., Hajianfar, G., Rezaeijo, S.M., Ghaemi, M., Rahmim, A.: Advanced automatic segmentation of tumors and survival prediction in head and neck cancer. In: Andrearczyk, V., Oreiller, V., Hatt, M., Depeursinge, A. (eds.) HECKTOR 2021. LNCS, vol. 13209, pp. 202–210. Springer, Cham (2022). https://doi.org/10.1007/978-3-030-98253-9_19
3. Singh, Y., Bhatia, P.K., Sangwan, O.: A review of studies on machine learning techniques. Int. J. Comput. Sci. Secur. 1(1), 70–84 (2007)

4. Salmanpour, M.R., Shamsaei, M., Rahmim, A.: Feature selection and machine learning methods for optimal identification and prediction of subtypes in Parkinson's disease. Comput. Methods Programs Biomed. **206**, 106131 (2021)
5. Salmanpour, M.R., et al.: Optimized machine learning methods for prediction of cognitive outcome in Parkinson's disease. Comput. Biol. Med. **111**, 103347 (2019)
6. Salmanpour, M.R., et al.: Machine learning methods for optimal prediction of motor outcome in Parkinson's disease. Physica Med. **69**, 233–240 (2020)
7. Oreiller, V., et al.: Head and neck tumor segmentation in PET/CT: the HECKTOR challenge. Med. Image Anal. **77**, 102336 (2022)
8. Iddi, S., Li, D., Aisen, P.S., Rafii, M.S., Thompson, W.K., Donohue, M.C.: Predicting the course of Alzheimer's progression. Brain Inform. **6**(1), 1–18 (2019). https://doi.org/10.1186/s40708-019-0099-0
9. Vallieres, M., et al.: Radiomics strategies for risk assessment of tumour failure in head-and-neck cancer. Sci. Rep. **7**(1), 1–14 (2017)
10. Javanmardi, A., Hosseinzadeh, M., Hajianfar, G., Nabizadeh, A.H., Rezaeijo, S.M., Rahmim, A., Salmanpour, M.: Multi-modality fusion coupled with deep learning for improved outcome prediction in head and neck cancer. In: Book Multi-Modality Fusion Coupled with Deep Learning for Improved Outcome Prediction in Head and Neck Cancer, pp. 664–668. SPIE (2022)
11. Butowski, N.A.: Epidemiology and diagnosis of brain tumors. CONTINUUM: Lifelong Learn. Neurol. **21**(2), 301–313 (2015)
12. Kumari, N., Saxena, S.: Review of brain tumor segmentation and classification. In: Book Review of Brain Tumor Segmentation and Classification, pp. 1–6. IEEE (2018)
13. Daisne, J.-F., et al.: Tumor volume in pharyngolaryngeal squamous cell carcinoma: comparison at CT, MR imaging, and FDG PET and validation with surgical specimen. Radiology **233**(1), 93–100 (2004)
14. Rodrigues, R.S., et al.: Comparison of whole-body PET/CT, dedicated high-resolution head and neck PET/CT, and contrast-enhanced CT in preoperative staging of clinically M0 squamous cell carcinoma of the head and neck. J. Nucl. Med. **50**(8), 1205–1213 (2009)
15. Roh, J.-L., et al.: 2-[18F]-Fluoro-2-deoxy-D-glucose positron emission tomography as guidance for primary treatment in patients with advanced-stage resectable squamous cell carcinoma of the larynx and hypopharynx. Eur. J. Surg. Oncol. (EJSO) **33**(6), 790–795 (2007)
16. Eyassu, E., Young, M.: Nuclear medicine PET/CT head and neck cancer assessment, protocols, and interpretation. StatPearls [Internet] (2022). StatPearls Publishing
17. Taxak, N., Scholar, M.T., Singhal, S.: A Review of Image Fusion Methods
18. Wang, Q., Shen, Y., Jin, J.: Performance evaluation of image fusion techniques. Image Fusion: Algorithms Appl. **19**, 469–492 (2008)
19. Salmanpour, M.R., Hajianfar, G., Lv, W., Lu, L., Rahmim, A.: Multitask outcome prediction using hybrid machine learning and PET-CT fusion radiomics. In: Book Multitask Outcome Prediction using Hybrid Machine Learning and PET-CT Fusion Radiomics (2021). Soc. Nuclear. Med.
20. Zwanenburg, A., et al.: The image biomarker standardization initiative: standardized quantitative radiomics for high-throughput image-based phenotyping. Radiology **295**(2), 328–338 (2020)
21. Salmanpour, M.R., Shamsaei, M., Saberi, A., Hajianfar, G., Soltanian-Zadeh, H., Rahmim, A.: Robust identification of Parkinson's disease subtypes using radiomics and hybrid machine learning. Comput. Biol. Med. **129**, 104142 (2021)
22. Bank, D., Koenigstein, N., Giryes, R.: Autoencoders. arXiv preprint arXiv:2003.05991 (2020)
23. Rahmim, A., Zaidi, H.: PET versus SPECT: strengths, limitations and challenges. Nucl. Med. Commun. **29**(3), 193–207 (2008)

24. Wang, G., Huang, Z., Shen, H., Hu, Z.: The head and neck tumor segmentation in PET/CT based on multi-channel attention network. In: Andrearczyk, V., Oreiller, V., Hatt, M., Depeursinge, A. (eds.) Head and Neck Tumor Segmentation and Outcome Prediction. HECK-TOR 2021. Lecture Notes in Computer Science, vol. 13209, pp. 68–74. Springer, Cham (2022). https://doi.org/10.1007/978-3-030-98253-9_5

25. Xie, J., Peng, Y.: The head and neck tumor segmentation based on 3D U-Net. In: Andrearczyk, V., Oreiller, V., Hatt, M., Depeursinge, A. (eds.) HECKTOR 2021. LNCS, vol. 13209, pp. 92–98. Springer, Cham (2022). https://doi.org/10.1007/978-3-030-98253-9_8

26. Salmanpour, M.R., et al.: Advanced survival prediction in head and neck cancer using hybrid machine learning systems and radiomics features. In: Book Advanced Survival Prediction in Head and Neck Cancer Using Hybrid Machine Learning Systems and Radiomics Features, pp. 314–321. SPIE (2022)

27. Salmanpour, M.R., et al.: Prediction of TNM stage in head and neck cancer using hybrid machine learning systems and radiomics features. In: Book Prediction of TNM Stage in Head and Neck Cancer Using Hybrid Machine Learning Systems and Radiomics Features, pp. 648–653. SPIE (2022)

28. Salmanpour, M.R., et al.: Deep versus handcrafted tensor radiomics features: application to survival prediction in head and neck cancer. In: EANM (2022)

29. Tang, M., Zhang, Z., Cobzas, D., Jagersand, M., Jaremko, J.L.: Segmentation-by-detection: a cascade network for volumetric medical image segmentation. In: Book Segmentation-By-Detection: A Cascade Network for Volumetric Medical Image Segmentation, pp. 1356–1359. IEEE (2018)

30. De Jay, N., Papillon, S., Olsen, C., El-, N., Bontempi, G., Haibe-Kains, B.: mRMRe: an R package for parallelized mRMR ensemble feature selection. Bioinformatics 29(18), 2365–2368 (2013)

31. Ishwaran, H., Kogalur, U.B., Blackstone, E.H., Lauer, M.S.: Random survival forests. Ann. Appl. Stat. 2(3), 841–860 (2008)

Deep Learning and Radiomics Based PET/CT Image Feature Extraction from Auto Segmented Tumor Volumes for Recurrence-Free Survival Prediction in Oropharyngeal Cancer Patients

Baoqiang Ma[1](✉), Yan Li[1], Hung Chu[1,2,3], Wei Tang[4],
Luis Ricardo De la O Arévalo[1,2], Jiapan Guo[1,2], Peter van Ooijen[1,2], Stefan Both[1],
Johannes Albertus Langendijk[1], Lisanne V. van Dijk[1], and Nanna Maria Sijtsema[1]

[1] Department of Radiation Oncology, University of Groningen, University Medical Center
Groningen, Groningen, The Netherlands
b.ma@umcg.nl
[2] Machine Learning Lab, Data Science Center in Health (DASH), University of Groningen,
University Medical Center Groningen, Groningen, The Netherlands
[3] Center for Information Technology, University of Groningen, Groningen, The Netherlands
[4] Department of Neurology, University of Groningen, University Medical Center Groningen,
Groningen, The Netherlands

Abstract. Aim: The development and evaluation of deep learning (DL) and radiomics based models for recurrence-free survival (RFS) prediction in oropharyngeal squamous cell carcinoma (OPSCC) patients based on clinical features, positron emission tomography (PET) and computed tomography (CT) scans and GTV (Gross Tumor Volume) contours of primary tumors and pathological lymph nodes.

Methods: A DL auto-segmentation algorithm generated the GTV contours (task 1) that were used for imaging biomarkers (IBMs) extraction and as input for the DL model. Multivariable cox regression analysis was used to develop radiomics models based on clinical and IBMs features. Clinical features with a significant correlation with the endpoint in a univariable analysis were selected. The most promising IBMs were selected by forward selection in 1000 times bootstrap resampling in five-fold cross validation. To optimize the DL models, different combinations of clinical features, PET/CT imaging, GTV contours, the selected radiomics features and the radiomics model predictions were used as input. The combination with the best average performance in five-fold cross validation was taken as the final input for the DL model. The final prediction in the test set, was an ensemble average of the predictions from the five models for the different folds.

Results: The average C-index in the five-fold cross validation of the radiomics model and the DL model were 0.7069 and 0.7575, respectively. The radiomics and final DL models showed C-indexes of 0.6683 and 0.6455, respectively in the test set.

Conclusion: The radiomics model for recurrence free survival prediction based on clinical, GTV and CT image features showed the best predictive performance in the test set with a C-index of 0.6683.

B. Ma and Y. Li—These authors contributed equally.

Keywords: Recurrence-free survival · Outcome prediction · Deep learning · Radiomics · Oropharyngeal cancer · Head and neck cancer

Team: RT_UMCG

1 Introduction

Worldwide, head and neck cancer accounts for around 900,000 cases and approximately 400,000 deaths annually [1]. The main treatment options are surgery, radiation therapy, chemotherapy or a combination of those three. For OPSCC patients, human papillomavirus (HPV) positive patients have much higher 5-year survival rates than HPV negative patients (45%–50% vs. 75%–80%) [2]. Several studies have investigated treatment de-intensification strategies for HPV positive patients with promising results [3]. The development of reliable prognostic outcome prediction models could identify high versus low risk patients for specific outcomes such as recurrence or survival, thus selecting patients who may benefit from treatment intensification or de-intensification. Many clinical predictors such as HPV status, age and TNM-stage were demonstrated to be predictive for treatment outcome in HNC patients [4, 5].

Radiomics, a method that extracts a large number of features from medical images, has been applied in HNC outcome prediction. Prediction models with radiomics features showed comparable or better prediction performance than models based on clinical data [5, 6]. However, radiomics is limited in reproducibility and standardization due to its dependence on image acquisition parameters, image quality, manual segmentation and handcrafted features [7].

DL has shown good performance in various medical image analysis tasks such as image synthesis [8–10], segmentation [11, 12] and cancer outcome prediction [13]. Compared with radiomics features, DL could extract more representative and comprehensive image features using convolutional neural networks (CNNs) and it has shown promising outcome prediction performance in HNC patients [4, 14]. ResNet [15] is a widely used CNNs architecture in classification tasks and has shown its potential in time-to-event outcome prediction for HNC patients [14]. However, clinical implementation of DL is challenged by its need for a large training set of medical images, which is typically not available, and, the extracted features are not explainable so they are less acceptable than radiomics for doctors. In this study, we compared DL and radiomics methods for recurrence-free survival prediction based on pretreatment positron emission tomography (PET) and computed tomography (CT) scans, auto-segmented GTV contours and clinical data.

2 Method

2.1 Data and Preprocessing

HECKTOR 2022 [16] provided training and test sets. The training set consists of 489 OPSCC patients who received radiotherapy with or without chemotherapy/surgery in

7 centers (Table 1). For each patient a pretreatment FDG-PET and a low-dose non-contrast CT scan of the head and neck region was collected on a combined PET/CT scanner. Furthermore, contours of the primary Gross Tumor Volumes (GTVp) and nodal Gross Tumor Volumes (GTVn) were provided. The outcome endpoint was Recurrence-Free Survival (RFS) with events defined as local regional and distant recurrences and censoring all others, including deaths. Time-to-event is defined in days starting from the end of radiotherapy treatment. In the training set an event rate of 21% and a median RFS of 14 months were observed. The test set included 339 patients from three centers (patient number): MDA (200), USZ (101) and CHB (38). The same data was available for patients in the test set as for patients in the training set except the GTVp and GTVn contours and RFS endpoint.

Table 1. Patient numbers and the way to use them in five-fold.

	CHUM	CHUP	CHUS	CHUV	MDA	HGJ	HMR
Number	56	44	72	47	197	55	18
Training	Except Fold 1	Except Fold 2	Except Fold 3	Except Fold 4	Always	Except Fold 5	Always
Validation	Fold 1	Fold 2	Fold 3	Fold 4	No	Fold 5	No

A bounding box (size: 192 mm in three directions) including tumors and lymph nodes was generated for each patient according to the brain position that was automatically segmented in the original PET scans using the code provided in the HECKTOR 2021 challenge [16]. In detail, the cranial border of the bounding box is defined 30 mm cranial from the caudal border of the brain. This defines the position of the center of the bounding box in the craniocaudal direction. In the lateral direction, the center of the bounding box is defined in the center of the brain and in the ventrodorsally direction it is shifted 38 mm to dorsal with respect to the center of the brain. Then, PET, CT and GTV images in the bounding box volume were extracted and re-sampled to $1 \times 1 \times 1$ mm^3 with trilinear interpolation. PET and CT intensities were truncated outside the range of [0, 25] and [−200, 200], respectively. For model training, we used 5-fold cross-validation in which patients from HMR and MDA were always included in the training set together with patients from 4 of the 5 others centers (Table 1). The 5th center was included in the validation set. HMR and MDA were always included in the training set because HMR included 18 patients only, which was not enough to use it as a validation set and the test set included a relative large number of patients from MDA.

2.2 Radiomics Model

Radiomics Features Prepared. Because there were no GTV contours available in the final test set, we used the DL auto contouring method developed by our RT_UMCG team for task 1 to segment the GTV in both the training and test set. This auto-segmented GTV contour was also used for radiomics extraction. The GTV primary tumor (GTVp) and GTV lymph nodes (GTVn) contours were merged into one region of interest called GTV, dealing with issues some patients did not have a primary tumor. The second reason was that it avoided the problems due to errors in labeling the auto-segmented contours as GTVp or GTVn. The Pyradiomics package was used for radiomics extraction [17]. Pre-processing of the PET and CT images and the auto segmented GTV contours were used for radiomics calculation. The applied bin size for the PET and CT images were 0.25 [18] and 25 [19], respectively. Texture features were computed from each 3D directional matrix and averaged over the 3D directions. Shape related IBMs were extracted from the GTV contours, intensity features were obtained from the PET and CT images respectively, texture features were calculated from CT images only because of the limited spatial resolution of the PET images.

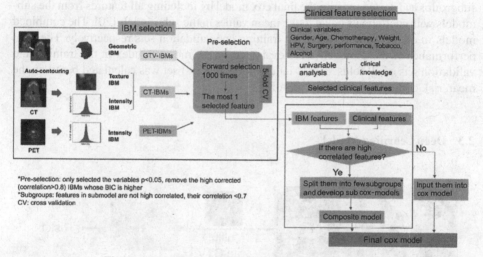

Fig. 1. The procedure of radiomics model development.

Model Development. The selection procedures of both clinical and IBM features can be seen in Fig. 1.

Clinical Features Selection. There were 9 clinical variables provided by HECKTOR Challenge, including gender, age, weight, chemotherapy, HPV status, surgery, performance status, tobacco and alcohol. Univariable analysis was performed in the complete training cohort for the clinical variables that were available for all patients: gender, age and chemotherapy. For the rest of the clinical variables, only the complete cases were used for univariable analysis. Selection of clinical variables was based on correlation with RFS in an univariable analysis and clinical relevance based on literature and expert knowledge.

IBM Features Selection. For IBM features, GTV-IBMs, PET-IBMs and CT-IBMs were analyzed separately since they were considered to give complementary information.

Pre-selection for those three types of IBMs was done separately by only focusing on the significant IBMs in univariable analysis and removing the high correlated IBMs (>0.8) with higher Bayesian information criterion (BIC). Considering the variances between centers, 5 folds were split as stated before. In each fold, forward selection was applied in 1000 bootstrap samples, and the criterion satisfied 2 conditions: BIC decreasing and the p value of the likelihood ratio test was significant when comparing the nested model with and without newly added feature. This procedure generated 5 frequency lists for each type of IBMs. The most often selected 1 feature (only one) in each fold for each type of IBM was counted and added to the final cox model if the count was three or larger.

When the correlation of the above selected IBM features was high (>0.8), we split them into different subgroups because the correlation would influence the coefficients of the correlated IBMs. After grouping, cox sub-models were developed for the different subgroups and combined into the final cox model by including all features from the sub-models with coefficients equal to the mean values in the sub-models [20]. The combined models in each fold were tested on training and validation sets respectively. The final performance of these models were evaluated by the mean C-index in the training and validation sets of all folds. The C-index in the final test set was calculated based on the mean risk of the 5 models from all folds.

2.3 Deep Learning Model

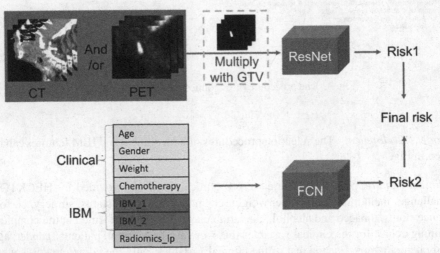

Fig. 2. The architecture of the DL model. IBM: two image biomarkers used in the radiomics models; FCN: fully connected network with three hidden layers with nodes: 64, 32, 16, respectively; Radiomics_lp: the linear predictor of radiomics models.

ResNet is chosen as the basic architecture of our model in this study. More specifically, we used 3D ResNet18 including 18 convolutional layers with the default settings from package MONAI (https://monai.io/). The basic architecture of our DL model is shown in Fig. 2. The input of the model could be any combination of CT, PET, GTV, clinical data, IBMs (the selected IBMs in our radiomics models) and the linear predictor of our radiomics model. Notably, the input of ResNet is the multiplication of the GTV mask with the bounding box volume of the CT and/or PET images in the case that the GTV is used otherwise the input is the bounding box volume of the CT and/or PET images. It was expected that using the GTV mask helps the model to focus on the most relevant volume, which is expected to be the tumor volume. The loss function used for model training was the negative likelihood loss [21] of the DeepSurv method which predicts the risk score of each patient.

Five models for the different folds were trained on a Tesla V100 32G GPU based on PyTorch 0.6.0 and MONAI 0.8.1 packages for a total 400 epochs using a SGD optimizer with a momentum of 0.99. Data augmentation included random flip in three directions with a probability of 0.5 (p = 0.5), random affine transformation (rotation degree: 7.5, scale range [0.93–1.07], translation range: 7 and p = 0.5) and random elastic distortion (rotation degree: 7.5, scale range [0.93–1.07], translation range: 7 and p = 0.2). Oversampling was used to select censored and uncensored patients with equal probability in the training process. The initial learning rate was 0.0002 and decreased by a factor of 0.2 at the 200th and 300th epochs. The training stopped when the validation C-index did not increase in 100 continuous epochs to avoid over-fitting and the model with the best validation C-index was saved for test. Finally, the predicted risk scores of the five models for the different folds were ensemble averaged to calculate the risk score in the test set. The scripts of radiomics and DL models are public in https://github.com/baoqiangmaUMCG/HECKTOR2022.

2.4 Model Evaluation

The C-index was used to evaluate the discriminative power of radiomics and DL models. Furthermore, the risk stratification ability of the models were analyzed by stratifying patients in the training and validation sets into high- (predicted risk > median risk) and low- risk groups (predicted risk ≤ median risk). Kaplan Meier curves for both risk groups were generated and the log-rank test was used to compare the RFS difference between the risk groups. A p-value < 0. 05 represents a significant difference.

3 Results

3.1 General Data Description

The clinical data included treatment center, gender, age, tobacco, alcohol consumption, performance status (Zubrod), HPV status and treatment (surgery and/or chemotherapy in addition to the radiotherapy). Only center, age, gender and chemotherapy were available for all patients. The other clinical parameters were only available for a limited group of patient (see Table 2 for the overview). Table 2 missing data count for clinical variables.

Table 2. Overview of number of patients with missing data count for the different clinical variables.

Variables	Gender	Age	Weight	Chemo
N missing (%)	0	0	6 (1)	0
Tobacco	Alcohol	Perf	Surgery	HPV
294 (60)	328 (67)	266 (54)	187 (38)	166 (34)

Chemo: Chemotherapy; Perf: Performance status; HPV: HPV status

3.2 Radiomics Model

Clinical and Radiomics Features. In univariable analysis, we found that only chemotherapy, weight, tobacco, HPV and performance status were significant features in all training sets. Even though chemotherapy was significant, it was not selected for the final model because the variation in chemotherapy between the different centers could be large [22]. Tobacco and performance status data were missing for many patients (for 294 and 266 patients, respectively), which made them not suitable for selection in the final model. Therefore, only weight and HPV status were selected for the final model. For weight, there were only 6 patients with missing data, which were imputed by the mean value of complete weight based on gender. For HPV status, there were 166 patients with missing data. And we chose a simple method due to time limitations that considered the missing data as another category, so the label of HPV status of 0, nan, 1 was converted to 0, 1, 2.

In total, 15 GTV-IBMs, 17 PET-IBMs and 92 CT-IBMs were extracted. After univariable analysis and removing the high correlated features, there were 6 GTV-IBMs, 5 PET-IBMs and 14 CT-IBMs left. The frequency lists (Table 3) of GTV-IBMs showed that shape_surfacearea was always the most selected IBM in all 5 folds and it was selected for the final model. The glrlm_RunLengthNonUniformity was counted 4 times as the most selected feature in 5 folds in the CT-IBMs frequency lists and was also selected for the final model. The frequency lists of PET-IBMs demonstrated that the 10th percentile and maximum SUV values were counted 2 times as most selected feature in 5 folds, which is less than the limit of 3 and means that they are less stable features than the CT features in the training cohort. Therefore they were not selected for the final model.

Table 3. Frequency lists of the IBMs candidates after 1000 times bootstrapping in 5 folds. Most selected features are indicated in bold.

A. Frequency lists of GTV-IBMs

GTV IBMs	Frequency in fold 1	Frequency in fold 2	Frequency in fold 3	Frequency in fold 4	Frequency in fold 5
Surface Area	779	883	774	678	741
LAL	58	18	55	117	38
Sphericity	90	57	33	151	163
MAL	14	10	30	15	23
M2DDC	12	10	76	2	3
M2DDS	22	5	15	11	8

B. Frequency lists of CT-IBMs

CT IBMs	Frequency in fold 1	Frequency in fold 2	Frequency in fold 3	Frequency in fold 4	Frequency in fold 5
Glrlm_ RLNU	457	253	384	299	347
Maximum	82	81	237	120	192
Glszm_ SZNU	46	75	33	30	72
Glszm_ZE	38	293	34	186	29
Range	50	36	35	44	83
Gldm_ DNUN	142	126	30	173	96
Glcm_JE	16	4	3	20	9
Ngtdm_Busyness	45	62	47	32	58
Glrlm_ LRLGLE	48	5	2	33	32
Ngtdm_Strength	32	20	124	13	43
Glcm_IV	7	1	24	18	10
Ngtdm_Complexity	9	5	8	7	4
Gldm_DE	4	3	2	4	3
Gldm_ GLNU	10	25	25	7	8

C. Frequency lists of PET-IBMs

PET IBMs	Frequency in fold 1	Frequency in fold 2	Frequency in fold 3	Frequency in fold 4	Frequency in fold 5
Maximum	313	630	198	594	320
10Percentile	445	122	256	170	437
Median	23	63	18	56	67
Uniformity	74	126	487	110	81

(continued)

Table 3. (*continued*)

C. Frequency lists of PET-IBMs

PET IBMs	Frequency in fold 1	Frequency in fold 2	Frequency in fold 3	Frequency in fold 4	Frequency in fold 5
Energy	70	38	20	27	54

LAL: Least Axis Length; MAL: Minor Axis Length;M2DDC: Maximum 2D Diameter Column; M2DDS: Maximum 2D Diameter Slice
Glrlm: Gray Level Run Length Matrix; Glszm: Gray Level Size Zone Matrix; Gldm: Gray Level Dependence Matrix; Glcm: Gray Level Co-occurrence Matrix; Ngtdm: Neighbouring Gray Tone Difference Matrix; RLNU: RunLengthNonUniformity; SZNU: SizeZoneNonUniformity; ZE: ZoneEntropy; DNUN: DependenceNonUniformityNormalized; JE: JointEnergy LRL-GLE: LongRunLowGrayLevelEmphasis; IV: InverseVariance; DE: DependenceEntropy; GLNU: GrayLevelNonUniformity

Final Radiomics Model. The final models contained the clinical variables: weight and HPV status and the IBM features shape_surfacearea of the GTV contour and glrlm_RunLengthNonUniformity of the CT values. These 4 features were used to develop a cox model for each fold. The resulting models and their performance in each fold can be found in Table 4. A similar performance with a C-index around 0.69 in the training sets was observed. The C-index in the validation set showed a larger variation from 0.63 (fold 2) to 0.75 (fold 5). The ensemble of the predictions of these 5 folds were used to predict the RFS in the final test set, where a C-index of 0.6683 was achieved.

The KM curves of high and low risk groups in cross-validation are shown in Fig. 3. Significant differences ($p < 0.05$) between the high and low risk groups were observed in all training sets while no significant differences were seen in the validation sets.

Table 4. The C-indexes and coefficients of each model in the training and validations sets of the 5 folds

	Coefficients of variables in final model						C-index	
	Weight	HPV status = 0	HPV status = 1	HPV status = 2	Surface Area	Glrlm_RLNU	In training set	In validation set
Fold 1	−0.0138	0	−0.5643	−0.7969	1.0723	0.1094	0.6836	0.7382
Fold 2	−0.0149	0	−0.8684	−0.9939	1.0833	0.1254	0.6872	0.6322
Fold 3	−0.0121	0	−0.6686	−0.8282	1.1149	0.1129	0.6899	0.6817
Fold 4	−0.0167	0	−0.7553	−0.8659	1.0897	0.1118	0.6949	0.7331
Fold 5	−0.0134	0	−0.5642	−0.9160	1.0942	0.1076	0.6892	0.7495
Average	/	/	/	/	/	/	0.6890	0.7069

RLNU: RunLengthNonUniformity

3.3 Deep Learning Model

The average C-index values achieved by the DL models in the validation sets in the 5 different folds are summarized in Table 5. The model which uses PET, GTV, clinical data (age, gender, weight and chemotherapy), 2 IBMs and the linear predictor of the radiomics model achieved the highest average validation C-index of 0.7575.

Table 5. The average C-indexes of the 5 folds in the validation sets

PET	PET Clinical	CT PET Clinical	PET GTV	PET GTV Clinical	CT PET GTV Clinical	PET GTV Clinical IMBs	PET GTV Clinical IMBs Radiomics_lp
0.7188	0.7278	0.7212	0.7393	0.7398	0.7408	0.7409	0.7575

Radiomics_lp: the linear predictor of the radiomics model

We further evaluated the risk stratification of the DL models in the training and validation sets as shown in the Kaplan Meier curves of Fig. 4. Significant differences between high and low risk groups were observed in all training sets and in the validation sets in fold 1 and fold 3. The C-index of the DL model in the final test set was 0.6455 which is much lower than the C-index of 0.7575 in the validation sets.

4 Discussion

The final radiomics model including weight, HPV status, shape_surfacearea of the GTV contour and glrlm_RunLengthNonUniformity of CT values showed the best performance in the final test set with a C-index of 0.6683. This result was comparable to the C-indexes observed in the five training and validation folds (Table 4). This demonstrates the good general applicability of the model in data from 9 different centers. Furthermore, it is an indication that no overfitting of the model occurred. Another indication for the general applicability of the model was the small difference in model coefficients in the five different folds except the missing HPV (HPV = 1) with coefficients from 0.56 to 0.87 (Table 4). The clinical and radiomics features in the model allow some clinical interpretation. OPSCC patients with a lower weight and patients which are HPV negative have more risk of recurrence after radiotherapy, which is consistent with clinical experience. Patients with a larger surface area of the GTV and a larger heterogeneity of the GTV have a higher risk of recurrence. A larger surface area is related to a larger volume of the tumor and pathological lymph nodes and a more irregular shape, which are known risk factors [5]. Also a larger tumor heterogeneity was shown to be related to worse treatment outcome in other studies [6].

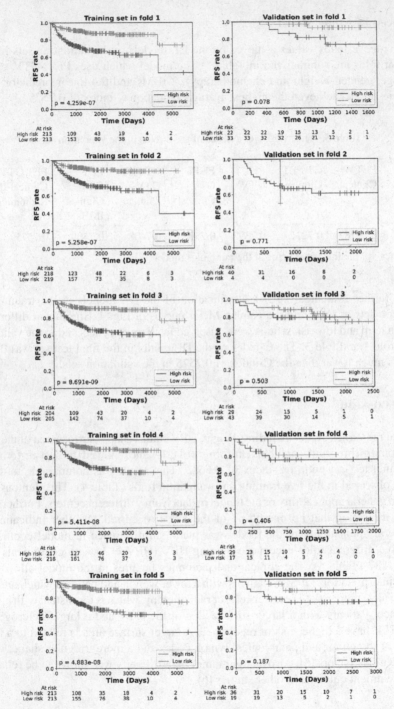

Fig. 3. KM curves of high and low risk groups stratified by the radiomics models.

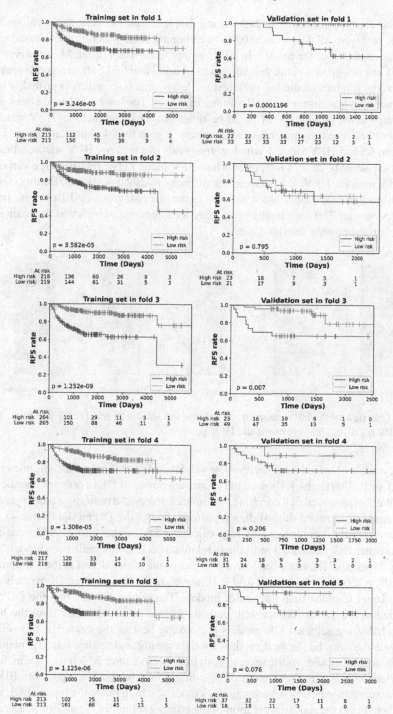

Fig. 4. KM curves of high and low risk groups stratified by the DL models.

In the risk stratification analysis using Kaplan Meier curves the p-values of the log rank test were all less than 0.05 in the training sets, which demonstrated a good stratification ability of the models in all 5 training folds. The Kaplan Meier curves of the high and low risk groups in the validation sets of fold 1 and fold 5 show a good separation, however, the p values of the log rank test are larger than 0.05 which is probably related to the limited cohort size and small number of events. The lower separation between the risk groups in the validation set could be related to the differences in HPV status (see Fig. 5) between the training and validation sets. E.g. in fold 4 almost all patients in the validation set had HPV value 1 which corresponds to a missing HPV status, whereas in the training set only 28% of the patients had HPV status 1. The relative large variation in the coefficients for HPV $= 1$ between the different folds supports this. Furthermore, the validation set of fold 2 only has 4 patients in the low risk group and 1 of those patients had an event after 750 days leading to a large RFS decrease, which was the main reason of the overlap between the risk groups.

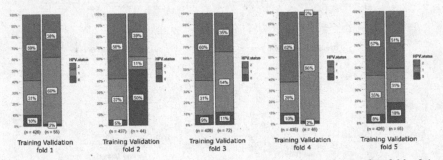

Fig. 5. Overview of the HPV status in the training and validation sets for the five folds. 0,1 and 2 are HPV negative, HPV unknown and HPV positive.

DL models obtained a higher average validation C-index than the radiomics models (0.7575 vs. 0.7069), which showed the promising power of DL in outcome prediction of HNC. We also observed that combing radiomics features and radiomics model prediction in DL models improved the validation performance of the DL model (Table 4), which may promote the combination of radiomics and DL in the future. Additionally, DL showed better risk stratification in the validation sets than the radiomics models (Fig. 4 vs. Fig. 3) because DL achieved lower p-values of the log-rank test in the validation sets especially $p < 0.05$ in fold 1 and fold 3. However, in the final test set, DL models did not show a larger C-index than radiomics models. The reason may be that the DL models overfitted in the validation set with patients from a different center than the training set. Probably we selected DL models performing best in the center of the validation set, which may not be the best models for the training and testing set. The majority of patients in training and testing sets are both from MDA center which was not included in the validation sets. Another reason of over-fitting may be that we input both IBMs and the linear predictor belonging to radiomics models, which made the input of DL models redundant. From the risk stratification of the training set in Fig. 3 and Fig. 4, DL models showed worse risk stratification than radiomics models, which also demonstrated that the selected DL models performed not good enough in the training set. Thus, splitting the patient cohort in training and validation sets with similar characteristics may be a better way to improve the DL performance in the test set.

Due to the lack of GTV contours delineated by physicians, only auto segmented GTV contours were used in this study. The limited performance of the auto-segmentation algorithm (Dice score of 0.65 in the final test set of task 1) could have influenced the results. Another limitation of the work was that only limited types of imputation methods were tested for the clinical variables with much missing data. It is expected that more advanced imputation methods could result in better prediction models, for example, it was shown in literature that the HPV status could be predicted using radiomics features from the CT images [23]. Even though the current model already showed good general applicability in data from 9 different centers, a more extensive validation in larger datasets from more different centers will be necessary before the models can be applied in the clinic.

5 Conclusion

Radiomics and deep learning models based on PET and CT image features combined with clinical features were developed for recurrence-free survival prediction in OPSCC patients. Models showed good predictive performance and risk stratification in the validation sets with C-indexes of 0.7069 for the radiomics models and 0.7575 for the DL models, respectively. In the final test, the performance of the radiomics model was better than that of the DL model with a C-index of 0.6683.

References

1. World Health Organization: Global cancer observatory. International agency for research on cancer. World Health Organization (2020)
2. O'Sullivan, B., et al.: Development and validation of a staging system for HPV-related oropharyngeal cancer by the International Collaboration on Oropharyngeal cancer Network for Staging (ICON-S): a multicentre cohort study. Lancet Oncol. 17(4) (2016)
3. Cramer, J.D., Burtness, B., Le, Q.T., Ferris, R.L.: The changing therapeutic landscape of head and neck cancer. Nat. Rev. Clin. Oncol. 16(11) (2019)
4. Ma, B., et al.: Self-supervised multi-modality image feature extraction for the progression free survival prediction in head and neck cancer. In: Andrearczyk, V., Oreiller, V., Hatt, M., Depeursinge, A. (eds.) HECKTOR 2021. LNCS, vol. 13209, pp. 308–317. Springer, Cham (2022). https://doi.org/10.1007/978-3-030-98253-9_29
5. Zhai, T.T., et al.: The prognostic value of CT-based image-biomarkers for head and neck cancer patients treated with definitive (chemo-)radiation. Oral Oncol. 95 (2019)
6. Zhai, T.T., et al.: Improving the prediction of overall survival for head and neck cancer patients using image biomarkers in combination with clinical parameters. Radiother. Oncol. 124(2) (2017)
7. Bi, W.L., et al.: Artificial intelligence in cancer imaging: clinical challenges and applications. CA: Cancer J. Clin. (2019)
8. Ma, B., et al.: MRI image synthesis with dual discriminator adversarial learning and difficulty-aware attention mechanism for hippocampal subfields segmentation. Comput. Med. Imaging Graph. 86 (2020)

9. Zhao, Y., Ma, B., Jiang, P., Zeng, D., Wang, X., Li, S.: Prediction of Alzheimer's disease progression with multi-information generative adversarial network. IEEE J. Biomed. Heal. Inform. **25**(3) (2021)

10. Zhang, X., Kelkar, V.A., Granstedt, J., Li, H., Anastasio, M.A.: Impact of deep learning-based image super-resolution on binary signal detection. J. Med. Imaging **8**(06) (2021)

11. Zeng, D., Li, Q., Ma, B., Li,S.: Hippocampus segmentation for preterm and aging brains using 3D densely connected fully convolutional networks. IEEE Access **8** (2020)

12. Oreiller, V., et al.: Head and neck tumor segmentation in PET/CT: the HECKTOR challenge. Med. Image Anal. **77**, 102336 (2022)

13. Diamant, A., Chatterjee, A., Vallières, M., Shenouda, G., Seuntjens, J.: Deep learning in head & neck cancer outcome prediction. Sci. Rep. **9**(1) (2019)

14. Wang, Y., et al.: Deep learning based time-to-event analysis with PET, CT and joint PET/CT for head and neck cancer prognosis. Comput. Methods Programs Biomed. 106948 (2022)

15. He, K., Zhang, X., Ren, S., Sun, J.: Deep residual learning for image recognition. In: Proceedings of the IEEE Computer Society Conference on Computer Vision and Pattern Recognition, vol. 2016-December (2016)

16. Andrearczyk, V., et al.: Overview of the HECKTOR challenge at MICCAI 2021: automatic head and neck tumor segmentation and outcome prediction in PET/CT images. In: Andrearczyk, V., Oreiller, V., Hatt, M., Depeursinge, A. (eds.) HECKTOR 2021. LNCS, vol. 13209, pp. 1–37. Springer, Cham (2022). https://doi.org/10.1007/978-3-030-98253-9_1

17. Van Griethuysen, J.J.M., et al.: Computational radiomics system to decode the radiographic phenotype. Cancer Res. **77**(21) (2017)

18. van Dijk, L.V., et al.: 18F-FDG PET image biomarkers improve prediction of late radiation-induced xerostomia. Radiother. Oncol. **126**(1), 89–95 (2018)

19. van Dijk, L.V., et al.: CT image biomarkers to improve patient-specific prediction of radiation-induced xerostomia and sticky saliva. Radiother. Oncol. **122**(2), 185–191 (2017)

20. Van den Bosch, L., et al.: Key challenges in normal tissue complication probability model development and validation: towards a comprehensive strategy. Radiother. Oncol. **148** (2020)

21. Katzman, J.L., Shaham, U., Cloninger, A., Bates, J., Jiang, T., Kluger, Y.: DeepSurv: personalized treatment recommender system using a Cox proportional hazards deep neural network. BMC Med. Res. Methodol. **18**(1) (2018)

22. Chamberlain, C., Owen-Smith, A., Donovan, J., Hollingworth, W.: A systematic review of geographical variation in access to chemotherapy. BMC Cancer **16**(1) (2015)

23. Leijenaar, R.T.H., et al.: Development and validation of a radiomic signature to predict HPV (p16) status from standard CT imaging: a multicenter study. Br. J. Radiol. **91**(1086), 1–8 (2018)

Author Index

V. Andrearczyk et al. (Eds.): HECKTOR 2022, LNCS 13626, pp. 255–257, 2023.
https://doi.org/10.1007/978-3-031-27420-6

Printed in the United States
by Baker & Taylor Publisher Services

Printed in the United States
by Baker & Taylor Publisher Services